25 WALKS

FIFE

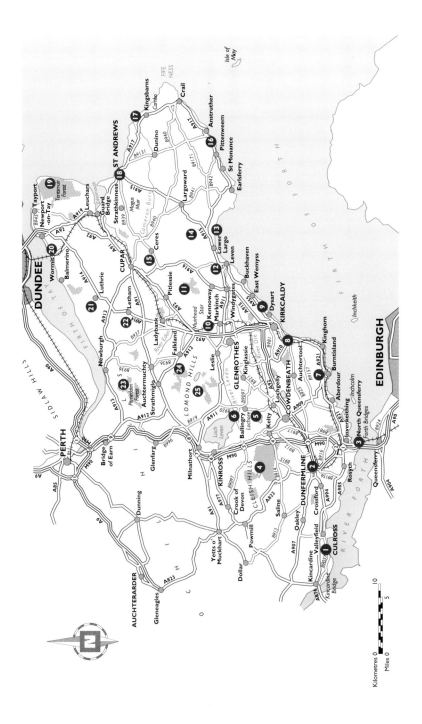

25 WALKS

FIFE

Hamish Brown

Series Editor: Roger Smith

FIFE
REGIONAL COUNCIL

EDINBURGH:HMSO

Applications for reproduction should be made to HMSO

Acknowledgements

Many people gave help and advice in the compilation of this selection of walks. I would especially like to thank Alison Irvine, Margaret Ramsey and others of the Fife Ranger Service, John Burrow, Manager of Vane Farm RSPB Reserve, the representatives and custodians of the National Trust for Scotland properties, Scottish Natural Heritage and Forest Enterprise staff, and the many farmers and local walkers – who were invariably proud of their Fife heritage.

HMSO acknowledges the financial assistance of Fife Regional Council in the preparation of this book

British Library Cataloguing in Publication Data

A catalogue record for this book is available from the British Library

Cover Illustration: Crail Harbour.
All photographs by the author.

Maps prepared by HMSO Cartographic Centre.

ISBN 0 11 495219 1

CONTENTS

USEFUL INFORMATION

The length of each walk is given in kilometres and miles, but within the text measurements are metric for simplicity. The walks are described in detail and are supported by accompanying maps (study them before you start the walk), so there is little likelihood of getting lost, but if you want a back-up you will find the 1:25 000 Pathfinder Ordnance Survey maps on sale locally.

Every care has been taken to make the descriptions and maps as accurate as possible, but the author and publishers can accept no responsibility for errors, however caused. The countryside is always changing and there will inevitably be alterations to some aspects of these walks as time goes by. The publishers and author would be happy to receive comments and suggested alterations for future editions of the book.

In the text the word 'path' is used for a purely pedestrian route, 'track' for private, unsurfaced but motorable routes (such as farm and forestry tracks) and 'road' for a tarred, public way.

A reminder about country behaviour: farm gates must be closed when used, farm buildings and machinery left alone, dogs kept on leads near livestock, growing crops avoided, matches not tossed aside or litter dropped. Routes described may or may not be rights-of-way. Farmers are usually friendly folk who work long hours to maintain the land we enjoy. When you meet, be helpful and friendly too!

The weather in Fife can be changeable but enjoys an east coast freshness and relatively low rainfall. If cloud covers the hills, walks lower down or on the coast can be rewarding. When you may be out for several hours water/wind-proof garments and strong footwear such as boots are advised, in most cases elsewhere shoes/trainers would suffice.

METRIC MEASUREMENTS

At the beginning of each walk, the distance is given in miles and kilometres. Within the text, all measurements are metric for simplicity (and indeed our Ordnance Survey maps are now all metric). However, it was felt that a conversion table might be useful to those readers who, like the author, still tend to think in Imperial terms.

The basic statistic to remember is that one kilometre is five-eighths of a mile. Half a mile is equivalent to 800 metres and a quarter-mile is 400 metres. Below that distance, yards and metres are little different in practical terms.

km	miles
1	0.625
1.6	1
2	1.25
3	1.875
3.2	2
4	2.5
4.8	3
5	3.125
6	3.75
6.4	4
7	4.375
8	5
9	5.625
10	6.25
16	10

Warning sign – a particular problem in Fife.

INTRODUCTION

Fife is a 'Kingdom' which, once discovered, will always delight: a green and varied landscape, a coastline which lives up to its royal description of 'fringed with gold', surprising hills with views out of all proportion to their altitude, fascinating historic towns, castles and prehistoric sites – Fife has it all.

There is plenty of variety in the length and physical demands of this selection of Fife walks but most are well within the capabilities of ordinary walkers rather than any super-fit Munro-bagger. Many are circular routes with shorter versions available, linear routes on the coast have bus/train services to return to the start, and many routes would be enjoyed on family outings – something for everyone – but do allow plenty of time as there is much to see and do.

Several of the coastal walks are described more fully in my book *The Fife Coast* (Mainstream 1994), the Regional Council has issued good books on Fife's castles and archaeological heritage, and both the Buildings of Scotland (J Gifford) and RIAS (G. Pride) series have superb illustrated guides to buildings and monuments. Worth reading are TG Snoddy: *Afoot in Fife* (1950) and *Tween Forth and Tay* (1966). A national award scheme in 1993 named Fife's Regional Council as the most environmentally-friendly in Scotland. Several walks are in the Fife Regional Park, several take in National Trust for Scotland properties (so have further descriptive reading available) while some give a surprisingly remote feeling and both the Crail - St Andrews coast walk and the round of the Lomonds are full and challenging days..

The top visitor venues in Fife are the two sea centres, the Scottish Deer Centre and the Kirkcaldy Museum and Art Gallery. Three of these are on walks and the Deer Centre, on the A91 near Cupar, can easily be added after any of the walks in central or north Fife. Attractions near walks are briefly indicated. It must be confessed that a few walks on the edge of Fife stray into Kinross-shire or Perthshire, and one walk enters from the Lothians. I apologise for this poaching!

Unlike walking in the Highlands, many of the walks are excellent in winter when cold, crisp days can make walking a delight. Some consider a place like Vane Farm best in winter because of its seasonal influx of birds, which could also apply to the coast. Fife is a surprisingly hilly county but sea or lochs are constantly in view. Spring can be vibrantly green, early summer rich and the autumn colours magnificent, while July and August often disappoint. June is my favourite month: most places to visit are open, the landscape is patched with yellows of oilseed rape and whin, the stirring skylarks reel over the hills. Welcome to the Kingdom of Fife.

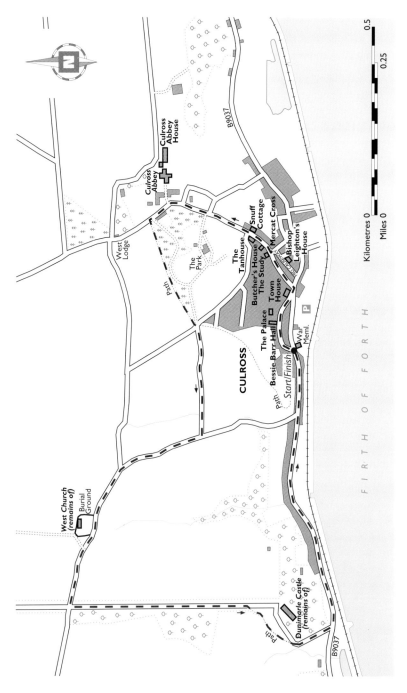

Culross Abbey House

Culross Abbey

West Lodge

The Park

Path

B9037

Snuff Cottage

The Tanhouse

Butcher's House

The Study

Mercat Cross

Bishop Leighton's House

Town House

The Palace

Bessie Barr Hall

War Meml.

P

Start/Finish

Path

CULROSS

West Church (remains of)

Burial Ground

Dunimarle Castle (remains of)

B9037

Path

FIRTH OF FORTH

N

Kilometres 0   0.25   0.5

Miles 0   0.25

HISTORIC CULROSS

Culross is only a small town lying on the Forth Estuary but it is a unique 17th-century survival and well worth exploring, a sanitised step back and walk about in the period when stone houses were first developing. The narrow cobbled streets, the tight houses with their red pantiles and crowstep gables have often been used as a film set. Early on our walk we pass the Town House (NTS) where a short historical video and guide books can fill out this brief introduction.

From the time of the monks, coal was a major industry, and cheap fuel beside the sea brought salt panning to Culross. (In 1633 there were 50 pans, so the town must have been a smoky hole.) Ships returning from Holland brought the red pantiles as ballast. Some of the first NTS purchases in the 1930s were made here: the Study cost £150, nine other houses £168. Longannet power station, prominent up-river, is still coal-fired.

The car park has display boards, anchors, the curved shape of an ice house, there is a toilet, war memorial and children's play area.

Cut down the lane immediately past the building close to the road with a 1681 lintel date and turn right for two NTS showpieces. Bessie Barr's malthouse is the building with the fore-stair up its gable (unusual) and, next to it, is the gem of the ochre-washed Palace, built by Sir George Bruce, part in 1597, the rest in 1611. There are rare tempera painted ceilings and period furnishings. Recently restored, it is again open to visitors. Note the windows with their mix of glass and shutters. The Town House (tolbooth), with its clock tower, shows Dutch influences. In 1975 when town councils ceased to exist it was handed over to the NTS as an interpretive centre.

Turn up the next street, the Back Causeway, cobbled and with a clear 'croon o the caussey' (crown of the

INFORMATION

Distance: 3½ km (2½ miles).

Start and finish: Main car park, west end of Culross, on B9037 between Forth Bridge and Kincardine Bridge.

Terrain: Pavements and paths; no special footwear needed.

Toilets: At the car park and some sites, cafes.

Refreshments: Several cafes and inns in Culross.

Opening hours: *Palace,* Easter, May-Sept., 1100–1700; *Town House,* 1100–1300, 1400–1700; *Study,* 1300–1600. Tel: (01383) 880359.

Culross Abbey ruins and Church.

Houses on Back Causeway.

causeway), the larger stones where the gentry walked clear of the margin rubbish. The narrow tower ahead is the Study and the corbelled room at the top is thought to be where the 17th-century Bishop Leighton worked. Across the street a date on a gable, 1577, is the earliest in the town. The mercat cross goes back to 1588. A merchant's house two up from the Study has a Greek inscription which translates 'God provides and will provide'. Before going up the Tanhouse Brae do walk down the Mid Causeway to see Bishop Leighton's house – and the coffee house may be a welcome break too.

On the Tanhouse Brae, left, a butcher's house has a cleaver and steel portrayed and, further on, right, the Snuff Cottage (1673) with a squeezed-in window, has the motto 'Wha wad ha thocht it' (who would have thought it?). This is in fact the first line of a couplet, the second line – 'Noses ha' bocht it' (noses have bought it) – appearing on the lintel of the snuff merchant's premises in Edinburgh. The Tanhouse is on the corner but you keep on up, aiming for the castle-like tower of the church, below which are some of the abbey ruins.

A church has stood here since early times. St Serf was here in the fifth century and a pupil of his, traditionally found new-born on the shore with his mother, was to become St Mungo, the saint of

Glasgow. An abbey was founded in 1217 but fell into
decay after the Reformation. The remoter West
Church was the parish church then but the
congregation successfully petitioned parliament in
1633 and the abbey became the church, the present
building being the original choir. A notable feature is
the Bruce Aisle, off the north transept, where the
family burial place has marble parental figures and a
kneeling row of three sons and six daughters, all in
period dress.

Outside, turn right to circle the graveyard. At a gate
Culross Abbey House is visible: 1608, very little later
than the Palace but years ahead in style and gracious
living. Cochrane, 10th Earl of Dundonald, and on a
par with Nelson as a naval genius, once lived here, a
figure largely ostracised by the establishments in
Britain.

Trade Stones abound: the hammermen's with crown
above the hammer, a butcher's with his cleavers, a
gardener's with crossed spade and rake, several with
maltsters' paddles and many with the sock and coulter
of ploughmen.

Ploughman's symbols in
Culross Churchyard.

Once out of the churchyard continue up the road
leading inland. Note the neat house on the right with
its 'eyes' of windows. Just past lodge gates follow the
footpath sign for the West Kirk. Cross a minor road,
then right at a T-junction and finally left at a Y-fork.
Some of the lintels in the church are older gravestones
with incised swords on them. There are many
hammermen stones and a big 4 is the symbol of a
merchant/trader. Plenty of skull and crossbones too.

Continue along the track till it crosses the tree-lined
but abandoned driveway of Donimarle Castle. Go
through the gate, left, and down this grassy avenue,
then, just before the last two big trees, veer right on a
green path (skirting the castle and its entrance) to
come out on the tarred drive which descends to the
coast road. Cross this towards the sea and there is a
footpath beside the railway which leads back
to Culross.

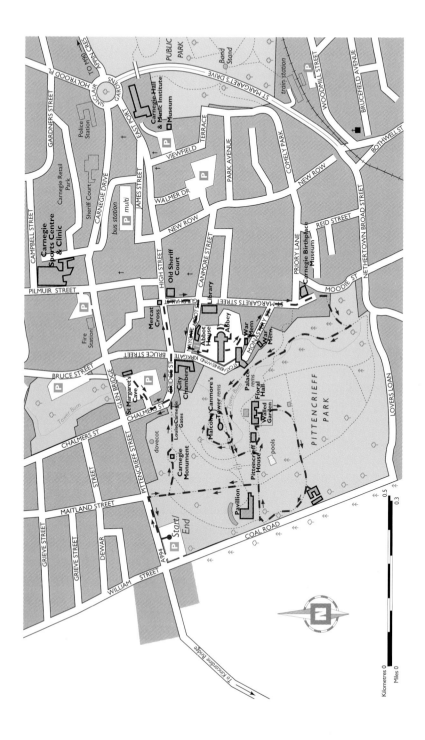

ROUND DUNFERMLINE TOWN

Dunfermline, ancient capital of Scotland, has one of Britain's finest parks so a walkabout gives a strange contrast of past and present, urban and rural, historic and modern. Free heritage walks take place every Saturday and Sunday from May to August; for details, telephone (01383) 724505/723638.

Enter Pittencrieff Park (locally called 'The Glen') from the gates on Pittencrieff Street then turn left at the first opportunity to pass the statue of Andrew Carnegie. Son of a weaver, he was banned from the Pittencrieff estate grounds as a boy but later, as the great American industrialist, he bought the estate – and promptly gifted it to the town. The huge gates beyond are a memorial to his wife. (To the left there's a small, battlemented doocot.)

In the formal garden of the Glen.

INFORMATION

Distance: 3 km (2 miles).

Start and finish: Car park, north-west corner of Pittencrieff Park (A994 exit from town).

Terrain: Town, park paths, no special footwear needed.

Toilets: Numerous.

Refreshments: Wide selection in the town and also the Pavilion in the park in summer.

Opening hours: *St Margaret's Cave,* daily (except Tues.) Easter-Sept. 1100–1700; *Abbot House Heritage Centre,* 1000–1700 daily, all year; *The Abbey,* Apr.-Sept. 0930–1700, Sun. 1400–1700. Oct.-Mar. 0930–1600, Sun. 1400–1600; *Pittencrieff House (Museum),* May-Oct. daily (closed Tues.), 1100–1700; *Pittencrief Park, Floral Hall and Aviary,* open daylight hours; *Carnegie Birthplace Museum,* Apr.-Oct., Mon.-Sat. 1100–1700, Sun. 1400–1700, Nov-Mar. daily 1400–1600; *Information Office,* 13 Maygate (01383) 720999.

Go through the gates and turn left, then cross and enter the car park to see St Margaret's Cave. It is actually *under* the car park. Margaret was an English princess, born in exile, who returned to the court of Edward the Confessor when she was eight. There she first met Malcolm, a refugee from Scotland, whose father, King Duncan, had been murdered by Macbeth. Fleeing again, when 20, her ship was driven into the Forth and she was helped by Malcolm, who then married her (1070). Devout Margaret had such an influence on Scotland that the Pope declared her a saint in 1249.

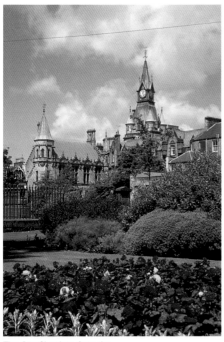

The City Chambers from the Abbey grounds.

Walk up the High Street, which becomes pedestrianised and has the old mercat cross facing the tall spire of what was built as a guildhall. The French baroque château-style building with the clock tower is the City Chambers.

Turn down Guildhall Street and right at the first crossroads. The impressive Central Library building is a Carnegie gift to the town. The tourist office is worth a visit for a town map or any queries before visiting the Abbot House next door. This, the oldest building in the town, has been restored and opened in 1995 as a Heritage Centre.

Continue along the Maygate until a gate gives access to the abbey grounds. Have a look at some of the 'trade' gravestones as you wander round leftwards to reach the end of the abbey and the site of St Margaret's shrine (only the marble base survives).

Rounding the building, you enter the church by the south transept. In a case by the door is displayed a bronze cast of the skull of Robert the Bruce, whose grave lies under the pulpit. Early last century, when the church was built on the ruins of the abbey choir, his tomb was found. (The rib cage had been cut open to remove the heart.) At the back of the church go through into the nave of the 12th-century abbey to see the magnificent pillars and Norman arches. Exit to the graveyard again.

The buildings across the graveyard, just shells, are partly monastic, partly royal palace (birthplace of Charles I) so turn left along the perimeter till a path leads down and right to pass the 1939–45 war memorial onto Monastery Street. Walk past the 1914–18 war memorial opposite and turn right to reach the Andrew Carnegie birthplace museum, a modest cottage (extended) with some interesting exhibits.

Carnegie birthplace.

Down and across, a gate leads into the Glen. Swing right before the play area to follow the meanderings of the Tower Burn. Keep beside it even when led through *under* the double-arched bridge. At a small footbridge over the burn, however, turn off right, to twist up and out onto a road.

This was at one time the main town access from the west, hence the bridge on top of a bridge as the steep climb was modified. Immediately right is a Historic Scotland notice about Malcolm Canmore's tower which you now visit. Only foundations survive. Walk to, and past, the tower to descend to the squirrel-haunted double bridge. Peer over, then walk on, under a bridge to come out near the Pavilion. The ground falls away as more traditional parkland and is very attractive. (Tea room and toilets available.)

Turn left for Pittencrieff House, built about 1610 and renovated early this century by Sir Robert Lorimer. It houses exhibitions. Beyond it are the glass houses (Floral Hall) with the superbly-kept formal gardens lying in front. West, right across the park, is the aviary and animal house. Peacocks will be met with anywhere.

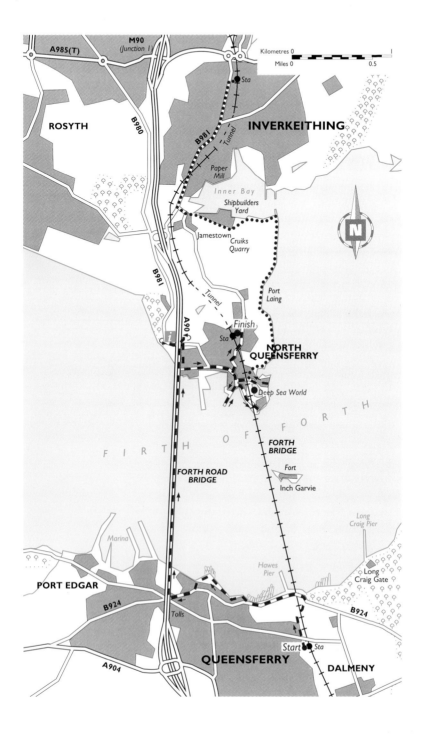

THE FORTH BRIDGE

This is a romantic way to enter Fife – walking across the Forth Road Bridge and, using the fabulous railway bridge, there is no difficulty returning to the start. Cars can be parked at Dalmeny Station, which lies above Queensferry.

Coming out from the station or its car park cross the road to take a footpath running parallel with the railway towards the Forth. Cross an old railway line by a footbridge then head half-right to drop down 120 steps (Jacob's Ladder) to the shore road. Turn left and walk along through Queensferry, a town of considerable charm and antiquity.

Sailings are made from Hawes Pier to Inchcolm (off Aberdour) with its historic abbey, and might be a ploy for a return visit. There are several cafes in the town and much of historical interest to see. The Hawes Inn is the setting of the opening scene in Robert Louis Stevenson's story *Kidnapped* and, until the road bridge was built, was where cars crossed the Forth by ferry. Note the sculpture of a seal and her pup.

The High Street (much of it 16th century) has houses on two stories (as in Chester) and black Black Castle is dated 1626 and has a secret stair. The old churchyard has some finely carved stones, one of a full-rigged ship, and many carvings of skulls. The Museum is interesting and the dominant tower along

A stone in the old Queensferry graveyard.

INFORMATION

Distance: 6 km (4 miles) or, to Inverkeithing, 10 km (6 miles).

Start and finish: Queensferry (Dalmeny Station); North Queensferry or Inverkeithing.

Terrain: Roads, bridge footway, coastal path, no special footwear required.

Toilets: In Queensferry, North Queensferry and Inverkeithing, also Visitor Centre at north end of the bridge.

Public transport: Frequent trains to Dalmeny from Edinburgh and Fife stations, and return trains from North Queensferry or Inverkeithing; enquiries: (0131) 556 2451.

Refreshments: Cafes, hotels and restaurants in the same settings.

Opening hours: *Queensferry Museum,* open all year, Thurs., Fri., Sat., Mon., 1000–1300 and 1415–1700, and Sun. 1400–1700; *Tourist Information Office* (north end of Forth Road Bridge), 1130–1500, 1730–2130; *Deep Sea World,* open 0900–1800, all year. Check return train times at Dalmeny at the start of the walk.

the High Street is the tolbooth with a commemorative well below it. There are covenanting memories, an old harbour and a one-time friary building. On a corner is a bakery/tea room which still has the old oven and such features preserved. You may welcome a cuppa here after exploring the town and before climbing up to gain the bridge.

The Forth Bridge from Queensferry.

Head up, passing Plewlandscroft to reach the bridge. Take the path signed *Cyclists only* (under the bridge), then turn left up the road to reach the complex of buildings servicing the bridge. No tolls or queuing for walkers! It is 2 km to reach the other side and, in gales, you can feel the whole bridge flexing. The towers are 160 m in height. The bridge has the lighting, drainage, road area (and servicing and finances) of a small town and is a memorable walk, however blasé we treat it as motorists. A plaque at the far side marks its opening in 1964 by the Queen, who then made the last ferry sailing. The ferry dates back to the time of Queen Margaret, wife of Malcolm Canmore; she gave her name to Queensferry.

Turn down the pedestrian steps. By passing under the bridge and up the other side you reach the Visitor Centre, Tourist Information Office and hotel/ restaurant area, which is a worthwhile diversion. Wend back and down into North Queensferry thereafter. There is an odd cluster of wells and, behind

The Friary-Inverkeithing.

them, the start of the coastal path. (The Napoleon's hat-shaped well aptly commemorates the Battle of Waterloo, 1815.) The coastal path leads round to Inverkeithing (¾ hour), an optional extra, from where a train can be taken back to the start.

It is worth wandering down to the pier and then round the bay to a viewing area right under the railway bridge – still one of the world's greatest engineering spectacles, the centenary of which, in 1990, was a great occasion. The bridge is floodlit in the evening. Inchgarvie is the island offshore, full of old wartime defences.

On the way back into town turn right at a footpath sign (*Helen Place* and *St James' Chapel*) and work through to reach the Deep Sea World complex – the world's biggest aquarium. The million gallon tank is the old flooded Battery Quarry and you travel along a moving walkway in a transparent tunnel . . . quite an experience! (Stone from the quarry was used on bridge foundations, canals, London streets and a Russian fort.)

The Forth Road Bridge.

A hydrofoil passenger ferry runs across to Queensferry which is an alternative way 'home' if the railway bridge is already well known, otherwise wend up the steep, twisting road from the wells to reach the station, or walk on to Inverkeithing, a much busier station, so more trains available, for travelling back to Dalmeny.

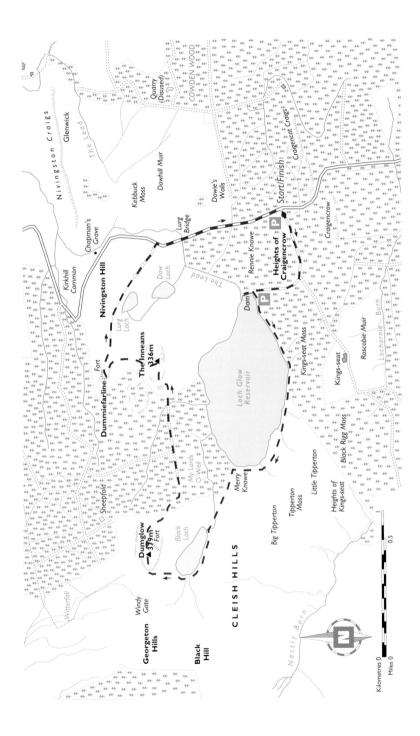

LOCH GLOW AND THE CLEISH HILLS

This is a hidden corner of upland country which has been progressively taken over by blanket afforestation, so the open spaciousness obtained from the 379 m summit comes as quite a surprise.

The start is a very small off-road car park, and easily passed. Look for the sign erected about fishing in Loch Glow and the Black Loch. Be early: Loch Glow is popular with fishermen! There are actually four lochs hidden away like navels in these bellyfolds of hills: Loch Glow, the Black Loch, Lurg Loch and Dow Loch; the walk circles them all.

During the fishing season it is also possible to drive down and park by Loch Glow, but for this walk the above is more useful. Loch Glow comes as quite a surprise and, if there is nobody about, can appear far bigger than it really is. The walk along the south shore is best made close to the water – the path along the bank above is much boggier. A few grouse survive on the moors but reeds and grass dominate over blaeberry and heather.

Sweep right round the loch to reach a wall at the far corner (Merry Knowe), then turn left to follow its line (which is also the old county boundary). There is a pretty view from a rise with the Black Loch backed by

INFORMATION

Distance: About 8 km (5 miles).

Start and finish: Small car park (099954), where forestry track leaves minor road. Leave the M90 at Junction 4 and head west, turning right after 2½ km onto a narrow road, signposted Cleish. The start is on the left, 3 km on. (Loch Glow sign at car park.) When leaving continue along this road and down by Nivingston, then turn right to regain the M90 at Junction 5.

Terrain: Mostly boggy moorland or grassy hills. Waterproof footwear advised.

Refreshments: None en route, but Nivingston Country Hotel Restaurant lies 2½ km to north. Tel: (01577) 850216.

Dumglow beyond Loch Glow.

the grassy Georgeton Hills, all with the complex knobbly character of the area's volcanic origins. The south side of the Black Loch is very wet to start but thereafter the going is firm and pleasant.

Cross the wall/fence at the end of the loch and start to climb up to the left (west) skyline of Dumglow. The saddle is Windy Gate, often an apt description. Cross a ruined wall and puff up by the fence. In the plain below, Aldie Castle is a landmark. This ascent gives a very steep section to reach Dumglow's summit trig point at 379 m. The view is quite fantastic with the Ochil Scarp and Devon Valley leading the eye away beyond the Wallace Monument to Ben Lomond, the Fife Lomonds rise beyond Loch Leven, Largo Law is seen over the shoulder of Benarty, the Bass Rock marks the far south-east with Edinburgh backed by the Pentland Hills, an astonishing panorama. The trig point sits on a prehistoric mound (a tree-coffin burial) which is circled by the walls of a fort. From the north rim you can look down on Cleish Castle, another typical tower house.

Head off eastwards (across several defensive wall lines) till the slope begins to fall away into forest. Ahead are the triple bumps of The Inneans, your next place to visit. Turn right to follow the fence until a basic stile is reached, cross and follow the path beyond. Over a brow this then leads down into the forest with a big break clearly seen going straight across to the three knobbly hills. Walk this break, cross the fence carefully, and climb the central of the three bumps, at 336 m almost as fine a viewpoint as Dumglow. The Lurg and Dow Lochs lie below your perch, the last of the day's collection.

The volcanic knobbliness of the Loch Glow Hills.

Due north, just right of the plantings which optimistically go right over the northern top of The Inneans, you can see another prominent knoll with a big triangular cairn on top. Walk over to this, picking up and using sheep trods. Dummiefarline (The Dummie, locally) is also a prehistoric defensive site. From it you see the motor road making its way down to Cleish; the parish church is prominent but the castle is hidden.

Head east, picking up a path outside the Lurg Loch bogs and flanking along Nivingston Hill. The path evolves into a track of sorts leading to a gate. Continue, bearing right a bit, to come out to the minor road at a gate/small sheepfold. A ten minute tramp along the road leads back to the start.

If forced to park by Loch Glow, after leaving Dummiefarline outflank The Inneans and keep to the drier, higher ground back to the dam and car park.

If you're a fisherman, Loch Glow offers trout fishing from 15 March – 6 Oct. Details/permits: Civil Service Sports Association (Tel: 01383 722123). If you're into challenging walks you could try the day's outing made by J. H. B. Bell and W. Omand in October 1931: Falkland (10.00), E. Lomond to W. Lomond (11.51), W. Bishop (13.00), over Benarty to Blairadam (15.21), Dumglow (16.50), Rumbling Bridge (19.00). Quite a hike!

MAY DAY ON THE CLEISH HILLS

Alone,
With the spiralled song of the lark;
Alone,
With the winds of the wild grey hill –
When the day is fired in the kiln of night
And eyes alive with the sights of the height.
Alone,
With the peewit's skeery screaming,
Alone,
Where the heart of the land lies still –
When the air grows cool and the heart glows bright,
When the step goes tired and the thoughts go light.
(Hamish Brown)

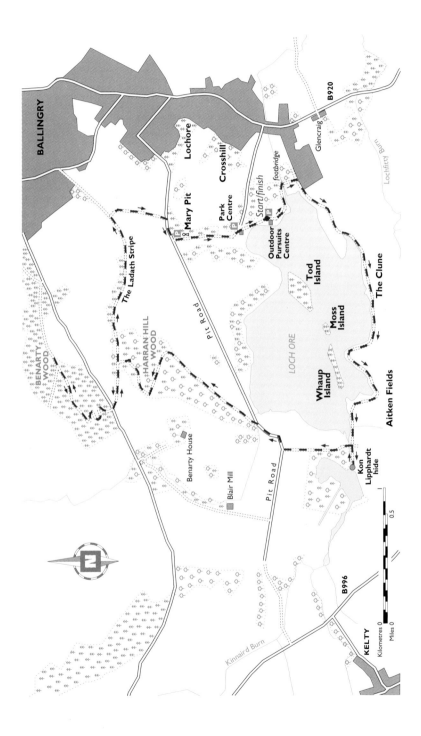

LOCH ORE CIRCUIT

Loch Ore has been re-created out of a landscape of mining dereliction and is now part of a popular Country Park offering sailing, wind surfing, canoeing, golf, horse riding, an adventure playground, birdwatching and various possibilities for walks, of which this is the longest and most rewarding, and one which would be excellent as a crisp winter outing or a floral spring day. Few people will be encountered on a surprisingly rural circuit.

Loch Ore lies surrounded by the one-time mining towns of Ballingry, Lochgelly, Cowdenbeath and Kelty and the only sign of this past is the monumental Mary Pit winding tower which stands in the park. The bings have gone, all is green or forested, with the park covering 460 hectares, of which 260 are Loch Ore itself, the re-emergence of a loch drained in the 18th century. The M90 (Junctions 3 or 5) offers quick and easy access to the park which is well signposted for motorists. (A wag in Lochgelly has touched-up the sign so it reads E ohore Country Park!)

The walk begins by wandering round the quiet southern shore of Loch Ore. In winter, swans, ducks, geese and gulls congregate on its waters. Ornithological sightings are on an update board in the centre for the enthusiast and there is a hide at the western corner, a nature reserve. Whaup Island is Scots for curlew island and they are not uncommon

INFORMATION

Distance: 8 km (5 miles).

Start and finish: Car park, Lochore Meadows Country Park.

Terrain: Good paths, grassy fields, woodland tracks. No special footwear required.

Toilets: In the Park Centre.

Refreshments: Cafeteria in the Park Centre. (Also information desk.)

Opening hours: *Lochore Meadows Country Park (and Centre)* open everyday, throughout the year, 0900–1700 in winter, 0900–2000 in summer (0800–2000 weekends). *Cafeteria,* 1100–1615 winter weekdays (Thurs.-Sun., closed 1330–1400), 1100–1730 and 1800–1930 all days, in summer.

Winter Day on Benarty with Loch Glow below.

Swans will be seen nesting and herons stalk the reeds where great crested grebes nest. The new plantings attract many migrants including willow and sedge warblers.

The walking path actually starts from the car park behind the Outdoor Pursuits Centre (signposted) and wiggles round to the outflow of the loch (R. Ore) which is crossed by a footbridge. At a T-junction turn right and the loch is soon rejoined. The volcanic ridges of The Clune have prehistoric hut circles and other evidence of ancient use. The path varies in character but, surprisingly, the second half is all over rich green cow-pastures, with odd tree belts (the Aitken Fields). Something like one million trees have been planted in the park, mostly alder, pine and birch. The west end of the loch is a nature reserve.

The prehistoric fort site on knoll of Dunmore above Lochore.

This is clearly signed. The path leads to a 'crossroads'; turn right, across a footbridge, the water this time flowing into the loch. By going straight ahead first, for a few minutes, you can reach the Kon Lipphardt hide, named after a local ornithologist who died the day before the hide was opened. The path, once over the bridge, runs straight along to meet the Pit Road, a tarred road, where you turn right. Through an S-bend, take the first gate on the left, signposted for Harran Hill Wood. (Going straight on completes a shorter route round the loch.)

The track climbs gently, then suddenly rears and twists up steeply onto and over Harran Hill, descending

Lochore Meadows Park with the visitor centre and boats for hire.

slightly and wending on to a gate onto a larger track, called the Avenue. The way back to complete the round is right, but it is worth going left first to Benarty Wood, to relish the big view that opens up. This is a steep haul but well-pathed or with steps. The track goes up to the tarred Hill Road (Ballingry - M90) and opposite rise the zig-zagging steps. The gradient eases and the path traverses along to join a forestry track, as far as you need go for the view. By turning left on the forest track, you can climb higher up and expand the view, but a return has still to be made down to Harran Hill Wood. Turning right on the forest track simply leads downhill to the Hill Road again.

Walking down the Avenue the verges are rich with wood anemones in spring (the woods are massed with wild hyacinth). The size of the Mary Pit winding tower is really noticed from up here. Just after an area of ruins, trees and some derelict allotments, go through the gate, right, into the field. Start down this and swing right to aim for the winding gear. A gate/stile leads back onto the Pit Road. Turn right, then left at the far edge of the car park to walk past the winding gear and an old shunting engine (or pug) to regain the start.

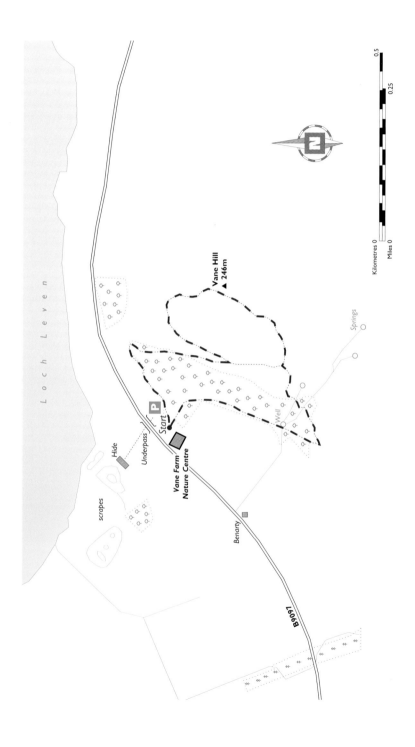

VANE FARM (Loch Leven)

Vane Farm was bought by the RSPB in 1967 to be an educational nature reserve and visitor centre and is a lively, interesting place with all the facilities for studying birds, a good shop, helpful staff and so on. There is a hide near the loch, reached by an underpass below the road, which should not be missed. This is a good place to explore in early winter (Oct-Dec) when the Icelandic geese are visiting, but the varied habitats make it a pleasant spot at any season and the nature trail gives superb views. Being a reserve, walkers are asked to keep to the trail. Keen birdwatchers can obtain up-to-date information in the centre.

The trail is well signposted and maintained. The top of Vane Hill is 246 m so the view is quite special. Loch Leven covers 14 square km and is backed by the Lomond Hills. The loch has romantic and historical associations, for Mary, Queen of Scots was imprisoned on Castle Island (summer boat trips from Kinross allow a visit). You may also see more than birds circling in the thermals over these Loch Leven hills for Portmoak (Scotlandwell) has the Scottish National Gliding Centre. Scotlandwell also has a very ancient canopied well and is only a few minutes drive from Vane Farm.

INFORMATION

Distance: 1½ km (1 mile). But allow 1 hour at least.

Start and finish: At the Centre, where there is a car park. Leaving the M90 at Junction 5, Vane Farm is well signposted and lies on the south side of Loch Leven (and is actually in Tayside, not Fife).

Terrain: Steep trail; stout footwear advised.

Toilets: At the Centre.

Refreshments: At the Centre, most weekends, otherwise Kinross has all the facilities of a town.

Opening hours: The Centre is open daily, 1000–1600 Jan.-Mar. and 1000–1700 Apr.-Dec., entry charge for non-RSPB members (dogs are not allowed on the reserve); *Loch Leven Castle,* Apr.-Sept., 0930–1830 weekdays, 1400–1830 Sun.

Looking to the Lomond Hills from Vane Farm.

Down to Vane Farm and to the far Ochils from Vane Hill.

The trail angles up and across by the regenerating woodlands, mostly birch and rowan, which have sprung up in the last twenty years, after sheep were removed from the reserve. Willow warblers, tree-pipits, redpoll and woodpeckers nest in the woods, and after the woods are left the path climbs steeply up through bracken, heather and blaeberry to the viewpoint of Vane Hill. Meadow pipits abound and many – poor things – are fostering cuckoos in their nests. Wheatears, whinchats, grouse and curlews may also be seen. (The higher crags on Benarty have nesting fulmars!) Some of the nest boxes are occupied by pipistrelle bats.

The viewpoint has an indicator. Kinross can be seen at the far left corner of the loch, straight ahead the Highland hills are glimpsed (often snow-touched well into summer), the Lomonds are bold above Kinnesswood on the right of the loch and, distantly, right of the cut, are the twin tops of Largo Law. The Bass Rock and North Berwick Law sit high beyond the estuary of the Forth. Beyond Benarty lie the Cleish Hills then a distant view west to the hills of Loch Lomond and the nearer, bold Ochil scarp.

The island nearest to the centre is St Serfs, on which you can see the ruins of an old priory. Ducks nest on the island. The loch level was lowered by 1.4 m in the 1830s when the long, straight cut was made to control the Leven's flow for the many mills downstream. The loch's vegetation has suffered over the years from the inflow of sewage, agricultural and industrial pollutants and has lost some bird species as a result. The lagoons or scrapes are man-made, to attract wading species and the winter visitors (geese by the thousand, whooper swans by the hundreds). Beware, people visiting Vane Farm simply to walk are apt to leave as nascent ornithologists!

Sunset on Vane Farm.

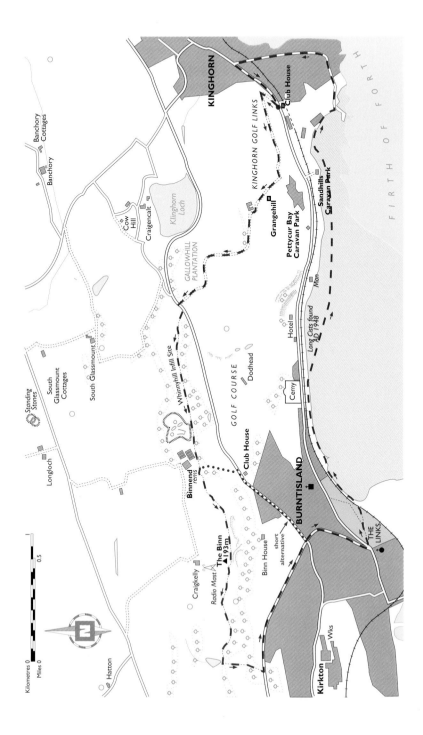

THE BINN (from Kinghorn)

The start is at the Kinghorn Golf Club building. If coming by bus ask to be put off there, otherwise walk west from the centre of Kinghorn (War Memorial by Alexander Carrick) and up the steps or drive facing the 1895 clubhouse. The track swings left at another clubhouse to rise steadily across the golf course. (Does hitting a pedestrian count as a penalty, I wonder?)

At the end of the golf course (Pettycur Bay Caravan Park on the left braes) the track swings inland, passing Grangehill House (left) and a farm beyond (right). Note the interesting 'window' on the south gable of the last building. All along this walk there are many rabbits and, hereabouts, black ones are not uncommon. Looking ahead, the Binn mast looks deceptively near.

As the track turns more inland the seaward setting is forgotten awhile; cows browse the rolling slopes and skylarks sing overhead. At another bend though the sea (Largo Bay) appears again, off right, with Kinghorn Loch, in its secretive hollow, below. The red slopes are waste alumina from the Burntisland aluminium works. When the old Kinghorn-Burntisland road is reached turn left and, after 30 m, cross carefully to a gate/stile beside the fenced-off area.

This long, gradually ascending track is the old road to the Binn village, at one time a thriving community of

INFORMATION

Distance: 10 km (6 miles).

Start and finish: Kinghorn Golf Club.

Terrain: Tracks, paths, steep grass and sandy beach. Strong footwear advisable.

Toilets: Kinghorn and Burntisland.

Refreshments: Kinghorn and Burntisland. Kingswood Hotel, between the two towns.

Opening hours: *Burntisland Museum:* during library hours; *Parish church,* by appointment (see Tourist Office for details).

The Binn seen over the Alcan works, Burntisland.

1000 people, owing its existence to the shale oil industry, now completely gone. (Under here and under the Forth and West Lothian lies a warren of old shale oil workings!) Halfway along the ascent, the water on the right is led across and down to a settling pool. Eye along the line of the ditch and, on the golf course, just over the road, you'll see a shelter made from half a boat. Aluminium and related products go into things as diverse as cosmetics, paint thickener and oven foil. The works will be seen from the top of the Binn.

A gate indicates the end of the long track up, and the ruins of the village lie on the right. Left, a gap in the wall and steps indicate a path down to the road if a shorter route is required, but two stiles and a couple of fields now lead to the top of the Binn, a viewpoint quite marvellously fine for its modest 193 m (622 ft). The cliffs fall steeply to Burntisland below and, being crumbly basalt, are propped up with concrete buttresses, avalanche barriers and a chain-mail of netting. There's a view indicator on the summit.

Behind the big mast is Benarty, right of it, the 'cones' of the Lomonds and, left, behind the billowing clouds created by Mossmorran are the lumps of the Cleish Hills – all walks described elsewhere. Forty miles off, Ben Chonzie is a Highland Munro, in the west the hills above Glasgow are clear.

Before leaving the summit, gauge the state of the tide. If it is up to the curve of the railway with no sands showing then the return to Kinghorn should be made along the main road. This passes the monument to Alexander III who went over the cliffs in 1286 on a riding accident when hurrying home to Kinghorn – a sad milestone event in Scottish history.

The route down keeps along the cliff edge, companioned by wind-contorted thorns, and over a grassy rise as high as the Binn itself. In the hollow beyond bear right and then down towards a lochan, picking up an old 'green road' round and down to the coot-noisy waters.

Go through a gate then turn left at a footpath sign to another gate and the trod leading across the field to a wood beyond. The path down through the wood to the road indicates old quarry workings here. Nature heals surprisingly quickly.

On Kinghorn's braes in Winter.

Turn left and walk down to the roundabout and on beyond: Cromwell Road (Cromwell won a vital battle a few miles west in 1651 and garrisoned Burntisland). This leads down to the Links (summer fair) with a cherub fountain at the town end. The Old Port marks a town gate back in the past. Note its triple, mottoed sundials and other adornments. There are several cafes and pubs off along right on Burntisland's High Street – which will probably be welcomed.

During library hours do look at the museum upstairs, laid out as a Victorian funfair. Ask too about the parish church (tourist office, 01592 872667, just round the corner from the library, in the Kirkgate) for it is a notable building and of historical interest. Once refreshed and Burntisland explored, head over the links to one of the tunnels under the railway to gain the promenade. This ends at a small, seasonal tearoom. If the tide is out miles of sand lead back to Kinghorn, if the tide is in, pass under the bridge and take the coast road. (At the Sandhills Caravan Park however, a footpath leads back down to Pettycur Bay.) Pettycur was once the main crossing point for travellers heading north. The road up from the harbour leads back to the start, and the High Street is well-served with cafes and pubs.

A shorter walk but enjoying the fine views from the top of the Binn can be had by a round walk from Burntisland: going out the old Kinghorn road to the golf club and the path opposite up to the Binn village and back as already described.

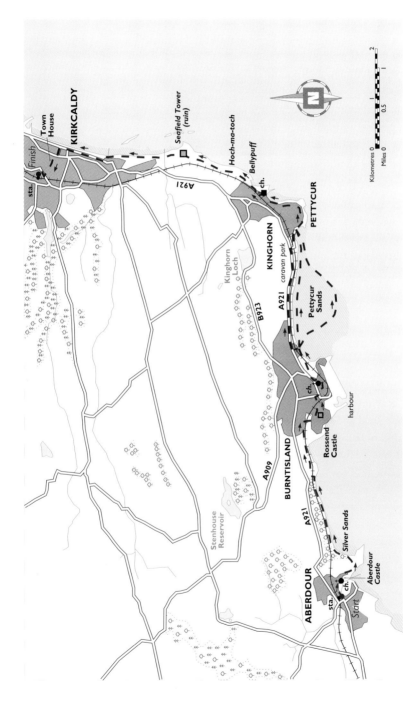

THE CASTLES COAST
(Aberdour to Kirkcaldy)

This is a varied and quite demanding walk so, while looking at sites and other interests, don't linger too long. The Kirkcaldy Museum and Art Gallery is popular so be there at least an hour before closing time. The station is immediately behind so returning to Aberdour is easy. If time becomes short you can finish at Kinghorn station. Aberdour won the Best Kept Station in Britain Award in 1990 and is a floral delight. Beside its entrance is the drive leading to Aberdour Castle (Historic Scotland) which is the first of several castles of note on the walk. The Earls of Morton owned Aberdour Castle, one of them having his head cut off by the Maiden (guillotine) and another imprisoning Mary, Queen of Scots in Loch Leven Castle. (His son was captured by Barbary pirates.) Only part of the building survives but there are several fine sundials and a notable beehive doocot.

From the castle go through to see the old church and its graveyard before heading on down to the Silver Sands and the well-made path along the coast to Burntisland, a woody walk by sea and railway until the huge complex of the aluminium works is reached. Turn right halfway along its perimeter and up the hill to reach Rossend Castle, a derelict tower which has been confidently restored and stands above Burntisland Harbour. It has its Mary, Queen of Scots association too. The hapless French courtier Pierre de Chastelard was found under the queen's bed (for a second time) and lost his head rather literally thereafter.

Burntisland Harbour and Rossend Castle.

INFORMATION

Distance: 16 km (10 miles).

Start and finish: Aberdour and Kirkcaldy are linked by both train and bus services but for reasons described, the train is recommended. If using a car, there is plenty of parking at stations.

Terrain: Mostly on paths but rough in places so stout footwear needed.

Toilets: In each of the towns passed and some of the sites mentioned and also the museum or station at the end.

Refreshments: Also found in the towns on the way (cafes and bar meals) and the museum at the end has a cafe.

Opening hours: *Aberdour Castle,* 0930–1800 weekdays, 1400–1800 Sun.; *Burntisland Museum,* Library hours; *Burntisland Church* details from Tourist Office; *Kirkcaldy Museum and Art Gallery,* 1030–1700 weekdays, 1400–1700 Sun.

Until the Forth Bridge was built a train ferry crossed to Burntisland and both Cromwell and the Romans thought it a good port. Work down to the High Street and along to the library building which houses a museum based on a Victorian fair (the fair continues to this day every summer). Round the corner (Kirkgate) is the information office, and if you walk along that way and turn left you find the gates of the parish church, well worth a visit.

Here, at a General Assembly, James VI set in motion the translation of the Bible that bears his name. There is no stained glass so the interior is light and airy, box pews survive and balconies bear heraldic and trade symbols while an outside stair let fishermen enter late if need be. There are several cafes at the east end of the High Street and The Links are then crossed to pass under the railway and back to walking the shore. See if you can spot the crocodiles on the cherub drinking fountain.

If the tide is out, you walk the rim of the huge Pettycur sands along to the little harbour of that name; if the tide is too far in then you will be forced to follow the road, which passes the monument to Alexander III, last of the Celtic kings, who rode over the cliffs here in the dark of a stormy night.

For centuries Pettycur was the main Forth estuary crossing for those heading up the east coast or to St Andrews, and this explains the 'Pettycur' on so many Fife milestones. Walk up Pettycur Road and turn off, right, at Doodells Lane which will lead you round to Kinghorn Bay where the unostentatious 1774 church is sited. The way on is blocked (there was once a thriving shipyard) so head up by the church, under the railway and then right at a play area onto the good path for Kirkcaldy.

If wanting refreshments go right up this road and left onto the High Street. (The continuation inland has some attractive 18th-century houses with marriage lintels, forestairs and pantiles.)

Hoch-ma-toch and Bellypuff are names attached to the first bay but note the rocks: lavas flowing down over limestone and shale levels as if they had only cooled yesterday. Wend along – and up – cross a wall and grassy slopes lead down to the stark, red ruin of Seafield Tower. There's a shoreline well and seals often haul out on the rocks just off-shore, and 'sing'.

Beyond the castle the path becomes a track and there is redevelopment taking place on what was once the site of the huge Seafield Colliery. There were bleach works, a rope factory, potteries, sweet factory and other industries too, all now swept away. At the grassy area keep to the edge of the sand and when a burn (the Tiel) bars progress just go upstream a bit and over a wall – and you are on the Kirkcaldy esplanade and will be made aware of why Kirkcaldy (pronounced Cur-coddy) is called the *lang toon*.

Seafield Tower.

Walk along the promenade until opposite the Comet site and turn inland, then turn right along the High Street. When it becomes pedestrianised swing left and walk straight on, passing the Tourist Information Office, then the fine Town House and Bus Station (right) and Sheriff Court (left) and crossing at a T-junction to reach what are war memorial gardens. The Museum and Art Gallery building lies at the top end of the gardens. Over on the right is the Adam Smith Centre with halls and a cafe.

The museum is a busy place and has several exhibitions each year besides its prize-winning local room, some famous paintings (Peploe, McCullough, Hornel among others), a shop and a cafe where you sit amidst show cases of glittering Wemyss Ware.

Adam Smith Centre, Kirkcaldy.

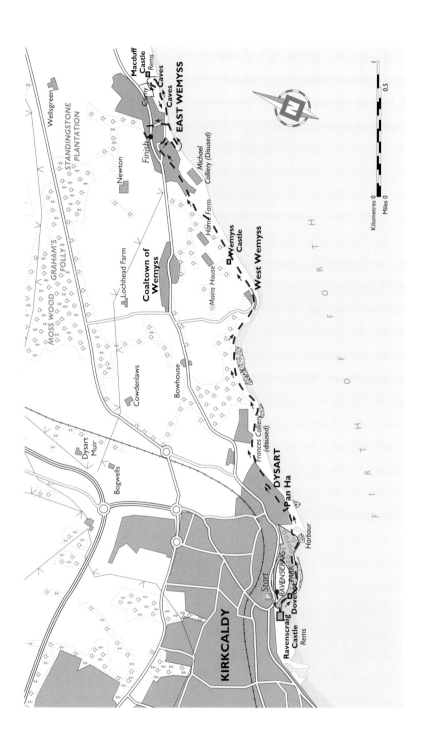

EAST WEMYSS

Macduff
Castle

Rems

Caves

Caves

Cany

Finish

Newton

Michael
Colliery (Disused)

Wellsgreen

STANDINGSTONE
PLANTATION

Home Farm

Wemyss
Castle

WEST WEMYSS

MOSS WOOD

GRAHAM'S
FOLLY

Lochhead Farm

Coaltown of
Wemyss

Mains House

Cowdenlaws

Bowhouse

Dysart
Muir

Bogwells

Frances Colliery
(disused)

DYSART

Pan Ha

Harbour

FIRTH OF FORTH

Start

P

RAVENSCRAIG
CASTLE PARK

Dovecot

Ravenscraig
Castle

Rems

KIRKCALDY

Kilometres 0

Miles 0

0.5

DYSART AND WEMYSS

This is another section of Fife's Castles Coast and, like Walk 8, gives plenty of things to see as well as enjoying the walking along by the Forth Estuary with its seascapes and wildlife. Fifty years ago it would have merited the name the Coal Coast.

Before leaving Ravenscraig Park follow the signs to Ravenscraig Castle, one of the more important architecturally, being the first built specifically to cope with the then new invention of cannons. This accounts for the thick walls (14 feet in places) and the sloping roof on the tower (so cannon balls skidded off). Ironically, James II who began the work of building the castle, was himself killed by an exploding cannon when besieging Roxburgh Castle in the Borders. The setting is splendid and Sir Walter Scott used it in his *Lay of the Last Minstrel*.

Once out of the castle head eastwards through the park, keeping close to the seaward side where there's a ridiculous zigzag wall which was built – spitefully – by a one-time landowner to stop workers walking from Dysart to Kirkcaldy along the shore. (He went bankrupt over gambling debts and the next owner presented the park to Kirkcaldy!) There's a well-preserved doocot (dovecot) as, anciently, fresh meat was rare in winter and every laird's castle had a doocot by it.

Eventually you will find yourself peering down directly onto picturesque Dysart, a real serendipity view, and a walkway and steps lead down to the harbour. The 'tower' is actually the old St Serf's Church, not a castle, though it could double as a defence, and the attractive white houses with the red pantiles line Pan Ha (*ha* = haugh = level area; now grassed), the pans being the ones used for making salt for which Dysart was once famous as it had the other vital ingredient to hand: coal.

Walk along Pan Ha and then up the steep Hie Gate to reach the town centre. The John McDouall Stuart

INFORMATION

Distance: 9 km (5½ miles).

Start and finish: Ravenscraig Park, Pathhead, Kirkcaldy (large car park) lies east of the main town and is the best start. The walk ends at East Wemyss whence coast buses will return you to Kirkcaldy (Pathhead).

Terrain: Mostly on paths or streets. Stout shoes adequate.

Toilets: In the towns, but none at Ravenscraig Castle.

Refreshments: Cafes in towns and the Belvedere Hotel, East Wemyss is strategically placed for walkers following today's route.

Opening hours: *John McDouall Stuart Museum,* Dysart, June-Aug., 1400–1700; *Wemyss Environmental Centre,* East Wemyss Primary School basement is open during school hours, 0900–1630.

Museum commemorates Dysart's most famous son. Born in 1815, he spent much of his life exploring Australia and was the first man to cross that great country from south to north. The Cross has a Victorian lamp marking it and a very photogenic tolbooth, with a 1576 date over the outside stair. Before you walk on, turn left up Cross Street to see the beautiful 16/17th-century houses opposite the end of the street. Heading east from the Cross the road climbs somewhat and when it swings left (council houses) bear off right to follow a path through and under the workings of the Frances Colliery.

This is still pumped dry in the hope it can be re-

opened. Sadly its redd (waste) was tipped directly into the sea, which caused all sorts of problems. A path has been built along the woody slopes beyond and eventually leads to a prow with a view to West Wemyss. Drop steeply down and along to this attractive quiet corner.

Ravenscraig Castle.

Among the restored buildings round the old harbour (now largely in-filled) the one-time miners' institute has been turned into the Belvedere Hotel (good bar meals, and old pictures of local interest). All signs of the mines have gone and many buildings in West Wemyss are empty, though you don't notice till you look carefully. Halfway along the Main Street there is a tall, slender tolbooth tower, with a pend (passage) through to the back. Tolbooths were customs points, gaols, town houses, etc. and, like doocots, Fife has the lion's share of those surviving. The swan symbol on the tolbooth is the heraldic device of the Wemyss family, whose castle dominates the shore shortly.

The town ends at St Adrian's Church whose graveyard has some stones of interest. One is made of coal and there seems to be a large ration of marble sculptures. Walking on eastwards, pass under the grim-looking Wemyss Castle (Wemyss comes from the

Gaelic name for *cave*) and the walk will end at the famous caves which lie beyond East Wemyss, itself indicated by the winding tower of the Michael Colliery.

Follow the path up to pass above the Michael Colliery. The miners' 'rows' have been attractively restored. The Michael had a disastrous fire in 1967 and, apart from the pumping gear, has been dismantled. Cut through and across the grassy area to Approach Row and from its end follow the road down to the shore again near the old parish church with the miniature soldier war memorial. The graveyard has some good trade-symbol stones.

Heading along the shore the caves appear on the left. They have suffered a century of vandalism and neglect yet contain more prehistoric carvings than the rest of Britain combined. Much has been lost due to roofs collapsing (the danger warnings are to be taken seriously) but it is safe to enter the Doo Cave, with its rows of rock-carved nesting boxes. Below Macduff Castle, the last on the Castles Coast, are more caves

Doo Cave – Wemyss Caves.

and, last, Jonathan's Cave with a grille across it to protect what are the best carvings. Those really interested should first go up to the Wemyss Environmental Centre in the basement of the primary school and obtain a key and booklets on the caves.

The centre is a good last port-of-call anyway and can be reached by going up the first curving street (heading back westwards) to the main road/bus route. The centre is immediately on the left and the enthusiastic staff will be delighted to chat to visitors and explain more about the caves. Ask too, about the local murder of the boy Michael Brown.

Coast buses will take you back towards Kirkcaldy, but ensure you get off at Pathhead as most buses do not actually pass Ravenscraig Castle.

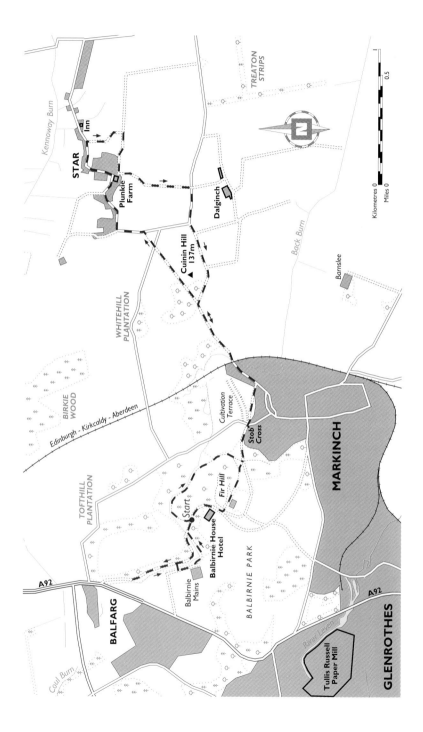

BALBIRNIE AND STAR

Balbirnie Park has many fine specimen and exotic trees (60 different species) and a wealth of rhododendrons collected from the Himalayas, so is worth visiting in early summer.

Markinch lies east of Glenrothes and is best reached from the A92, turning off the road facing the Tullis Russell Paper Mill entrance, which is signposted for Balbirnie Park. The entrance is on the left as Markinch is reached. Follow the drive past the mansion (Balbirnie House Hotel) and use the extensive woodland car park beyond. The prestigious hotel is open to non-residents and a meal, tea or refreshments there may be welcomed at the end of your explorations.

The 416-acre park is the old Balfour family estate which prospered in the 18th century with profits from improved agriculture and coal mining. From the road on past the mansion, fork right to the golfers' club house. Beyond, golden yews welcome you to the woodlands.

Stay on the right bank of the burn after an open area, the path climbing and giving views across to some soaring Wellingtonias. Ignore paths heading off right or left. When the woods end, there is a notice warning about subsidence, a legacy of the coal mining era. On entering woods again, turn left at the first opportunity

INFORMATION

Distance: 8 km
(5 miles)

Start and finish:
Balbirnie Park,
Markinch.

Terrain: Parkland and rural landscape, mostly on paths or tracks but waterproof footwear advisable.

Toilets: Balbirnie House Hotel car park, Plough Inn at Star.

Refreshments:
Balbirnie House Hotel; Plough Inn, Star.

Star.

and walk out to the East Lodge. Turn right, crossing to the narrow pavement.

Look back to see the ancient Stob Cross up on a bank, a rare example of a sanctuary cross connected with the parish church, the spire of which soon appears ahead. The site had an early church built by St Drostan, nephew of St Columba, and Markinch may have been the capital of Fife when it was one of the seven kingdoms of Pictland. The tower of the church is 13th century.

Not far past the Markinch town sign, the route goes through a gap in the wall and along past the obvious play area to reach a street where you turn left (50 m on the right there is a Victorian letterbox). Walk along and under an echoing railway bridge (the Edinburgh-Kirkcaldy-Aberdeen line) for the signposted path to Star. Left is a Victorian graveyard with a splendid array of yew trees. The Celtic cross for the Balfours of Balbirnie is the most interesting feature.

The Balfour cross in the North Hall graveyard.

The footpath for Star rises up behind the cottages, a steady pull. There's a long straight and just 40 m after the path bends, leave it to take the stone steps up left for a path through the woods of Cuinin Hill. At the top of the wood a kissing gate gives access to walk along the edge of a field. There is a sudden expanding of the view with Largo Law prominent and the village of Star straight ahead.

Cross the first road and a path opposite cuts the corner of the field to reach the road into Star. This is a charming village of cottages which has an almost unique rural atmosphere. Follow the twisting road through the village. Make a note of Plunkie Farm (right, red pantiles) where there's a footpath sign for Dalginch. You will follow this after visiting the far end of the village, joining it by another footpath starting 100 m beyond the post office (signed for Plunkie Farm/Treaton).

Right at the end of the village – just where walkers will welcome it – is the Plough Inn which has snacks, bar meals and a restaurant (open 12 noon-2.30 pm and 6 pm-midnight). Turn back and go up the Treaton/

Plunkie Farm track; it wiggles past some houses and then swings right to run along to Plunkie Farm. Turn left and follow the farm track which leads up to a minor road. Turn right and at the first sharp bend break off left on a farm track which is followed along (ignore track going left after 50 m) to tidy North Lodge. After that it becomes a footpath and joins the outward route from Markinch.

Return to Balbirnie Park by the East Lodge but walk straight on from it, ignoring other paths and the outward route. After a 'crossroads' take a path that angles up a grassy bank, right. It joins one contouring Fir Hill where there is a magnificent *Pinus nigra*, its arms bent to the ground, and backed by a row of monkey puzzle trees. Walk left above the old glasshouses to come out at the Balbirnie Craft Centre, created in the former stable block.

Some workshops are open to visitors and there's a fine art gallery in the corner of the yard. Coming out from the yard, turn right, then the first opening on the left (variegated holly) will lead back to Balbirnie House.

Before calling a stop there is one more short walk not to be missed. Continue along past the car park, over a bridge and on up the Balbirnie Burn (snowdrops in season under the trees), cross a gated transverse track and, at a grassy area, there is a prehistoric site of some interest. An interpretive board explains the attractive site. Walk back down to the two gates, turn right and cross the bridge to walk down the other bank of the burn which leads back to the start.

Snowdrops in Balbirnie Park.

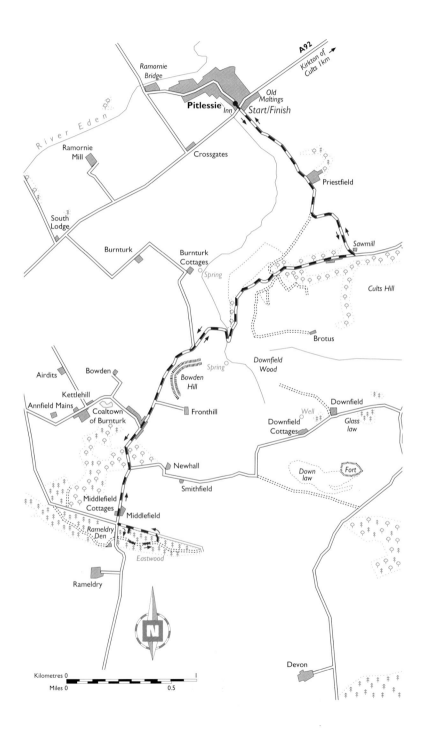

ABOVE THE HOWE OF FIFE

This is a gentle walk which gives sweeping views of Fife's heartland. It is something of a mystery tour too and is a good walk for a winter's morning, or an evening when the light can flood along the Howe of Fife. Pitlessie is a charming village forever associated with the painter Sir David Wilkie who grew up there and, in 1804, painted *Pitlessie Fair* (National Gallery of Scotland). The now derelict maltings point out the village's past history but Pitlessie is still a 'best-kept village' contender and typical of the Howe of Fife, as the great level sweep by the River Eden is called. With the river (and old Rossie Loch), drained in the 1740s and 1800s, the land is some of the richest in Scotland.

Cross the busy A92 from the historic inn to head up the good farm track to Priestfield, passing right of the farm to curl up onto the 'balcony road' as I've heard it called. Join it through old lime workings where there's a huge concrete lime kiln with a plaque 'Rebuilt 1937'. On the left is a sawmill, and in it you can see a surviving outcrop of limestone. Quarrying took place all along this 'balcony' of Cults Hill for centuries but has largely died out and now little is seen, so vigorous has been the regeneration of trees. Beside the cross-roads 1 km east,

The old coaching inn at Pitlessie.

however, there is still a busy brick and lime works, and more or less opposite where you come out onto the tarred road is the entrance to one quarry still in use. This whole upland area is called the Riggin o' Fife.

Unusual warning sign at Burnturk.

Turn right along the road. After about 500 m the trees finish and there are superb views over the Howe to the Lomonds and the northern rim of Fife. The road puts in a hairpin bend and then circles Bowden Hill (a prehistoric fort site) to reach the scattered hamlet of Burnturk. Beside its town sign is a triangle warning sign with a frog on it! (It is only displayed 'in season', so we have a picture of it.) The fuller, older name was Coaltown of Burnturk. Coal often lay conveniently near limestone measures.

Keep on ahead, passing a wood, then open country, and reach a wood clearly signed Eastwood, beside the Balmacolm junction. Turn up here, left, where there are three small parking spaces and then an arrow saying Entrance. To what I refuse to say, except that the footpath makes a circuit in the wood to come out further along the track and you then walk back to Burnturk and the 'Balcony' road. As this curiosity is not publicised I trust readers/walkers will leave others to be intrigued, and surprised in turn.

Over the Howe of Fife.

The walk back is along the outward route (there are no safe, practical alternatives) but you gain a whole new perspective and the views vary from hour to hour or minute to minute. Can I make a plea too for visitors, when driving, to explore some of these minor roads in Fife, both for the scenery and to see some of the sites and sights.

Back in Pitlessie, there is a post office/village shop and an inn to offer refreshments, and then a short historical addition can be made by car. Turn east onto the A92 (Cupar direction) and, after 1 km, turn off right at a sign for Kirkton of Cults. Cults church was re-dedicated in the 12th century so is an old site. The present church, with its ornate birdcage bellcote (which used old gravestones!) and outside stair, is late 18th century. Inside, it still has box pews and looks much as it must have a century ago. There are long ladles for taking up the collection. David Wilkie's father was the minister and there are many other family plaques, relatives going to India and Australia, very much part of that period's social history.

In the cemetery there are several Cochrane family stones, a family related to the Earls of Dundonald, noted at Culross. One stone commemorates a Cochrane couple happily married for 70 years! About 1 km to the east there is a tomb (not accessible) on a hill called Lady Mary's Wood where the Victorian builder of Crawford Priory mansion is buried. She was

Harvesting – Howe of Fife (above Pitlessie).

noted for running about the woods with the deer, starkers. Strictly speaking she rebuilt a reasonable lodge, turning it into an architectural hotch potch. The gutted shell still stands. Lastly, before leaving, note the fine early 18th-century doocot and the huge size of the one-time manse. In centuries past manses were often used like inns. No doubt pigeon pie appeared on the menu. The tiny building at the entrance is the Session House.

Both the walk, and wandering around here, will have given you some idea of past life in rural Fife. The folk museum at Ceres (Walk 15) would be a good follow-up day and the Hill of Tarvit (NTS) is one mansion that can be visited.

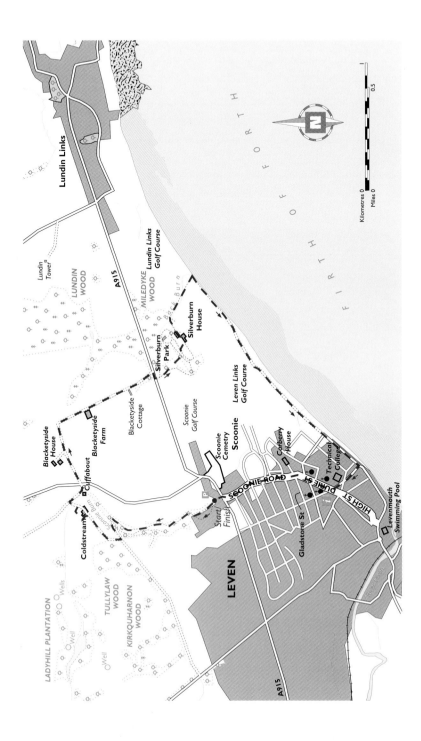

LETHAM GLEN AND SILVERBURN

This is an astonishingly varied walk and, with two places having animals and birds on show, offers much to children. The complete circuit, however, should not be undertaken by small children who would enjoy shorter variants or separate visits to Letham Glen and Silverburn. As a brisk *walk*, winter can often be a good season and the glen and woods are colourful in autumn.

Letham Glen gives a very pleasant garden and woodland walk. Head up the main drive as far as the Pets' Corner (aviaries, animal enclosures). Cross the bridge beside the doocot/rabbit warren and wend on upstream to the end of the glen. Cross the burn and take the highest path to angle up to a park entrance. (The option of a short circuit can be made by just walking down the glen again.)

Turn right on a farm track. When it swings right there is a grand view of Largo Law. Circle round the farm (Coldstream) and out by the track beyond. A small burn is crossed and a bigger burn, the Letham (Scoonie lower down) is followed downstream to a house called Cuffabout. Cross the bridge to reach the Leven-Cupar road and go straight across on the

INFORMATION

Distance: 6 km (4 miles).

Start and finish: Car park at Letham.

Terrain: Good paths, tracks and roads. No special footwear required.

Toilets: Letham Glen, Silverburn and Leven Promenade.

Refreshments: Leven; many facilities.

Opening hours: *Silverburn Mini Farm*, Mon.-Fri., 0830–1630, weekend afternoons; *Arts and Craft Centre*, weekend afternoons, Wed. afternoon (July, Aug.); *Letham Glen*, daylight hours; *Tourist Office:* Leven. Tel: (01333) 429464.

The doocot/rabbit house in Letham Glen.

Blacketyside cottages.

Blacketyside House private road. Past this attractive house the road turns sharp right to pass a big farm, then some fruit fields and attractive cottages, one of which is an Embroidery Workshop and Gallery. (Open Sun. 1300–1700, Tues.-Sat. 1000–1700. Tel: 01333 423985.)

Cross the busy A915 with care to reach Silverburn Park which has gardens, trails, specimen trees, a mini farm and an arts and crafts centre. Take the gate opposite the entrance (rather than the drive) and follow the footpath straight on to a T-junction with a bank beyond and views of the sea. Turn left, along a track which merges with another and, not far short of the open area by Silverburn House (the Arts and Crafts Centre), go through a sturdy gate on the right. This path descends through the attractive walled garden and a gate will allow an escape onto an avenue with the Mini Farm opposite.

This has a large collection of farm animals, various species of fowl (Scots Dumpy, Indian Runner) and a display of old machinery. Once out, turn right along past a row of red-roofed cottages and into the car park, right. From the far corner a path leads down to the sea, following the line of the Back Burn and a wall (the Mile Dyke), which separates the Leven and Lundin

Links golf courses. Once a year the clubs combine for a game, using half of each club's course and being entertained in the appropriate clubhouse at the half-way stage.

Walk westward along the edge of the golf course (or on the sands) to reach a caravan park. There is a path along beside it (on the landward side) and then you are back on the Leven promenade. Leven itself is the market town for this area of Fife so is well-provided with shops and refreshment choices. The Tourist Information Office and shop is in the Beehive building at the top end of the pedestrianised High Street and at the seaward end (near the huge chimney of the power station) is the very modern Levenmouth Swimming Pool and Sports Centre; tel: 01333 429866. You could fill a day quite easily between making the full walk, with its attractions, and enjoying the town.

To reach the town centre walk along the promenade and turn right at the roundabout. Walk up past the technical college. The road swings right (pedestrian lights), the way back to the start, while left is the pedestrianised High Street, starting at the Tourist Office with the 'Beehive' plaque above.

Silverburn – grooming time for the Shetland ponies.

Walking back to the start there's a mimic of the beehive up by a clock, left, but they forgot to show any bees! Durie Street, becoming Scoonie Road, has an incredible number of churches, the war memorial and, opposite Gladstone Street is Carberry House where the multi-faced sundial, which was once the Town's mercat cross, has been erected – a very unusual type. Back at the roundabout outside Letham Glen, do turn up the hill and go in to explore the Scoonie graveyard. There are some notable old stones and one with a beautifully carved ship. The best stones are at the Scoonie end of the site. Leven and Scoonie were once separate parishes and only joined up this century.

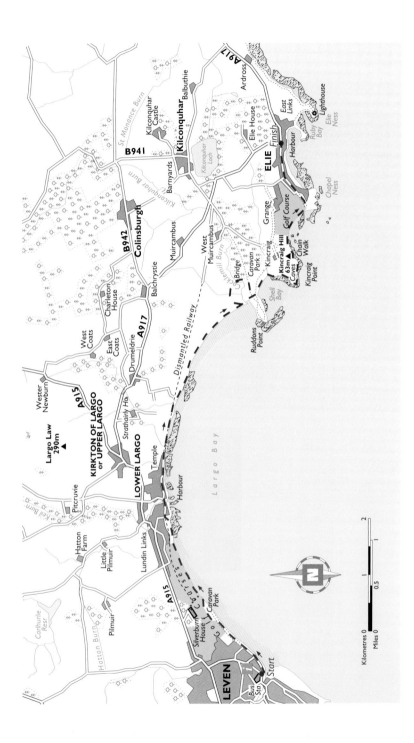

LARGO BAY (Leven to Elie)

This is a delightfully varied walk for it touches the first of the East Neuk towns with their special atmosphere yet also gives remote walking on dunes – and offers the energetic the oddity of the Chain Walk.

Wherever you park in Leven, head east along the promenade to the golf course and the Leven Beach Caravan Park. It can be passed on either side and the edge of the golf course followed. At the first burn (bridged) you can turn inland to visit Silverburn, a centre based on the old Russell estate with its flax mill, which is described on Walk 12.

Back at the Mile Dyke, swop to Lundin Links Golf Course and walk along its edge with a fine view of Largo Law (Walk 14) ahead. At the far end keep to the shore unless the tide dictates otherwise: there's a maze of rock features and waders in numbers often congregate in migration times and winter. We come to Lower Largo, with its river-mouth setting backed by the old railway viaduct and the Crusoe Hotel being the old granary building. The Crusoe, the Railway Inn and Harbour Cafe offer a welcome break.

Halfway through the old town you'll see why the Crusoe name appears here – Lower Largo was the birthplace of Alexander Selkirk, a vivacious lad who was eventually packed off to sea, did very well, and actually *asked* to be marooned on Juan Fernandez

INFORMATION

Distance: 16 km (10 miles).

Start and finish: Leven is a busy town for the central Fife coast and has a bus station facing the swimming pool complex at the mouth of the River Leven. Parking in big central car park, or on promenade. Walk finishes at Elie, whence bus back to Leven. (See note at end.)

Terrain: Streets of towns contrasting with dunes, seashore, coastal path and rock scrambling (optional, this last). Boots suggested.

Refreshments: Only at Lower Largo (Crusoe Hotel) and Elie finish (Golf Hotel, etc.), towns with a wide range of facilities.

Toilets: Leven promenade, Silverburn, Lower Largo, Elie.

Opening hours: *Silverburn.* Open office hours, weekdays, and afternoons at the weekend. (Walk 12 gives fuller details.) *Tourist Office:* Leven. Tel: (01333) 429464.

Lundin Links Golf Course.

Lower Largo.

Island, so bad a ship was he on. (It sank not long after!) I don't think he bargained for a four year and four months stint alone but, once back, Daniel Defoe made use of his experiences with the result we all know.

The road does a divergence round what was once the Cardy net factory, which still has a white mast topped with a salmon weather vane. Beyond this you are in Temple, as the east end of Lower Largo is called, possibly because of an early link with the Knights Templar. Where the town ends you begin the empty sweep of dunes backing fair Largo Bay.

The bay ends at the Cocklemill Burn which may come as a shock if the tide is in. (At low tide you can cross where it braids out on the sands.) The burn loops through a salt marsh but a well-hidden bridge lies upstream just below the St Ford Links wood (463009). From Ruddons Point, skirt outside a caravan site, along the well-named Shell Bay, with its big dunes.

Looking ahead, Kincraig Point looms and there are three distinct terraces of raised beaches showing what were one-time sea levels. The easy line wanders round to the point and then up these natural steps onto the top of the 63 m cliffs of Kincraig Point (radio mast and wartime defences still clear), keeps high for a while and then drops down to a golf course. The Chain Walk alternative keeps to the shore line and gives an energetic challenge – assuming the tide is out or falling.

The Chain Walk has carved steps and vertical or horizontal chains strung on the difficult sections of the cliffs so is demanding (and no place for dogs) but it does allow you to see several caves, some spectacular displays of columnar basalt and other features, besides being great fun in itself. If tackled, allow a full hour for this passage.

When either route chosen reaches the golf course follow along its seaward edge (or walk the warm gold sands) to reach a sunken track. Take this inland to reach Earlsferry, which runs on into Elie without any break, so it feels a long town.

On the Chain Walk.

The church, with its distinctive 1726 clock tower, marks the centre of Elie (it has three clock faces only as there were no buildings to the north when the clock was added in 1900), and there are some intriguing old stones on the wall by its door. The town is clean and fresh with golden sandy bays, a harbour much used by sailing and windsurfing enthusiasts and several excellent restaurants.

Buses for Leven stop opposite the church. Check times in advance if possible. (Kirkcaldy Bus Station, tel: 01592 642394 or St Andrews Bus Station, tel: 01334 74238, or look at the timetables displayed at Leven Bus Station before starting).

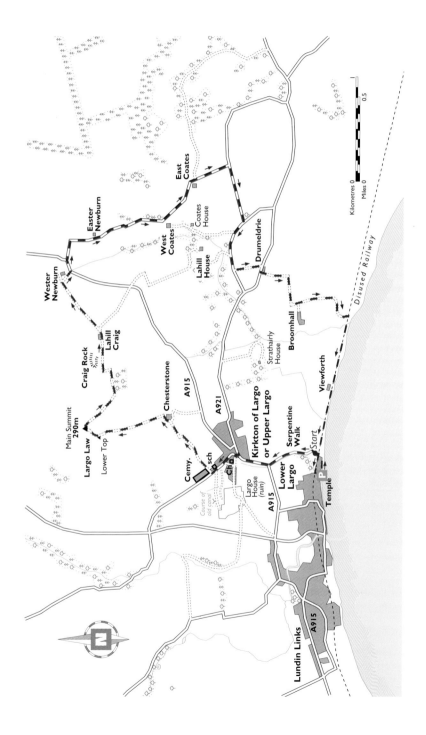

LARGO LAW

L ower Largo, on the coast, is linked with Lundin Links on the A915, so turn down for Lower Largo when indicated and head eastwards. The river mouth area has the Crusoe Hotel (originally a granary) and the high arches of the now defunct coastal railway, then, further along is the famous statue of Alexander Selkirk ('Robinson Crusoe') whose home was here. The single street is narrow and puts in a couple of tight-angle bends to come right to the sea's edge. The Temple car park is just on the left and a notice board gives something of the historical interest of the area.

Walk east for another 150 m where the start of the Serpentine Walk is signposted. This crosses the old railway line (now a walkway) and continues up woods owned by the Woodland Trust. Notice how trees are planed, west to east, by the prevailing winds.

The path comes out on the A915. Gates across the way led to Largo House, now derelict but with the fine Adam front still intact. Turn right, up the hill to Upper Largo. The inn sign of the *Yellow Carvel* commemorates one of Sir Andrew Wood's ships (Sir Andrew was a 15th century naval commander, a native of Largo, noted for his fights with English ships in the firth). Cross to take the first road left (just past the garage), which dips down and then heads off over the west flank of Largo Law. Before heading for the hill divert to the church, off left on its commanding site, to see a Pictish symbol stone just inside the

INFORMATION
Distance: 13 km (8 miles). Just up and down Largo Law is 6 km (4 miles).
Start and finish: Lower Largo, Temple car park.
Terrain: Mostly on roads and tracks but, on the law, very steep hillside where boots are advisable.
Toilets: Temple car park.
Refreshments: Lower Largo, at the end.

Lundin Links standing stones looking towards Largo Law.

gateway and, on the south side, a stone with elongated figures on one side and their heart-rending story on the back. Sir Andrew Wood is buried in the church.

Turn right coming out of the churchyard then take the first road left, which will get you back onto the hill road. Note the 'piebald' cottage, quite a local feature and often the black is simply the dark whinstone. Peer over the wall on the left before going further and you'll see a slight depression curving away towards a small white tower with a conical roof – all that is left of Sir Andrew Wood's castle. The depression marks a canal (the first ever cut in Scotland) which he had made by English prisoners so he could go to church in his admiral's barge.

Upper Largo – The Historic Church and Largo Law.

Continuing up the hill road, you come to the primary school and modern graveyard. Between them a kissing gate is the start of the recognised route up Largo Law. A field edge and then track leads up to Chesterstone Farm, passing a row of old cottages neatly converted into a house. Follow the track round the farm till, left, there is an obvious green lane heading straight for the Law. The final ascent is one of the steepest of all the walks described in this book and whin bushes allow no deviations low down. You arrive on a fine summit only to find there's a dip (fence with stile) and the real summit beyond, with a cairn and trig point.

The view is extensive: the sweep of Largo Bay along to the sprawl of Leven and the far Binn with its relay mast, seaward lie the May Island and Bass Rock, while

Lomonds, Ochils, Sidlaws and horizoned Highland Hills give a variety of mountain scenery. The summit area is surprisingly grassy and full of yellow violets, bedstraw, speedwell, buttercups and cuckoo flower.

Descend to the gap again, cross the stile and swing round and down the south top to reach the corner of a wood. Skirt the field edge beyond to reach the cottages below the Craig Rock and on by the track to Lahill Craig Farm (B&B with superb views). Keep straight on along the track to Wester Newburn and turn right there to reach the A915. Turn left and, almost at once, leave it, right, to follow a small, tarred country road which descends by the farms of Easter Newburn, West Coates and East Coates (B&B) to reach a T-junction, where you turn right.

Before doing so however, visit the ruined church on the left. The graveyard is in the care of the local Department of Recreation! In the south-west corner are buried members of the famous Lorimer family; professions given include architect, painter, professor of law, East Indian merchant and lawyer.

The road runs westward in a straight line then bears right at a junction, past the Coates House drive and up to a scattering of houses. At the far end of the hamlet, left, is a beech hedge and at its end a kissing gate allows you to walk down a field edge and driveway to the busy A921 at Drumeldrie. Cross and walk along (right) past the first buildings, then left onto a gravelly farm track which makes an angular, zigzag descent to the seashore. Be careful not to swing in to Broomhall – the way on is straight ahead, a track with a green strip down the centre. It becomes sandier and ends by the old railway. Partridges are common here.

Climb up onto the railway track, now a walkway, and follow it back to Temple/Lower Largo. Viewforth was once the site of salt pans and the posts for salmon nets can be seen by the bay. You can stay on the path till above the car park and then go down steps (made of sleepers) into it. The Temple name is thought to have connections with the Knights Templar.

Alexander Selkirk memorial – Lower Largo.

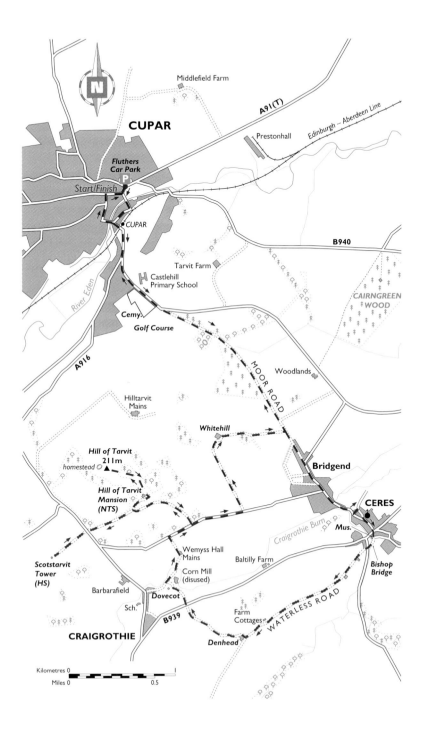

Middlefield Farm

A91(T)

Edinburgh – Aberdeen Line

CUPAR

Prestonhall

Fluthers Car Park

P

Start/Finish

CUPAR

B940

Tarvit Farm

H Castlehill Primary School

River Eden

CAIRNGREEN WOOD

Cemy.

Golf Course

A916

MOOR ROAD

Woodlands

Hilltarvit Mains

Whitehill

Hill of Tarvit 211m

homestead

Bridgend

CERES

Hill of Tarvit Mansion (NTS)

Craigrothie Burn

Mus.

Wemyss Hall Mains

Baltilly Farm

Scotstarvit Tower (HS)

Corn Mill (disused)

Bishop Bridge

Barbarafield

Dovecot

Farm Cottages

WATERLESS ROAD

Sch.

B939

CRAIGROTHIE

Denhead

Kilometres 0   1
Miles 0   0.5

THE MOOR ROAD, CERES AND HILL OF TARVIT

This is best done as an afternoon walk (pub lunch in Ceres perhaps) as the main visitor attractions are only open then. It is not really suitable for dogs or very small children.

Take care crossing the busy main road on leaving the car park, turn right and skirt the ornate war memorial. Fork left and round to a junction (Cupar station, left). Turn left, over the Edinburgh-Aberdeen line (the statue is to one of the line's backers) and walk on out of Cupar awhile before turning left onto the Ceres road. The town ends with the cemetery and golf course, then there is an obvious kissing gate into trees and a footpath sign for Ceres.

The Moor Road is typical of older routes in the pre-turnpike days when boggy valley bottoms were avoided. The shady climb is pleasant, there is some open ground on top and then the path runs down to meet the main road at Bridgend, as this part of Ceres is called.

Continue down into the village, an old, trim place and, opposite Meldrum's Hotel, turn up Curling Pond Road onto Kirk Brae and call in to see the Griselda Hill Pottery. This specialises in recreating the once-famous Wemyss ware and is now highly collectable in its own right. You can see the pre-glaze painting in progress from the showroom.

A little further on is the 1806 church, a large sandstone block with a landmark steeple. The site is ancient and the graveyard has some old stones and a flag-roofed mausoleum of 1616. See if you can find the wall monument to Rev. J C C Brown and the heart-breaking story it tells.

Drop down to rejoin Main Street (the village's shops) and on to a crossroads by the old coaching inn. Opposite is the toby jug figure of The Provost, which is actually the figure of a local divine.

INFORMATION

Distance: 13 km (8 miles).

Start and finish: Cupar, Fluthers car park, east end of town.

Terrain: Road, tracks and clear paths, with steep grass on Hill of Tarvit. No special footwear required.

Toilets: Several in Cupar, including Fluthers car park, Ceres centres, Hill of Tarvit mansion.

Refreshments: Wide range in Cupar, pub facilities in Ceres, tea-room, Hill of Tarvit mansion.

Opening hours: Ceres: *Fife Folk Museum,* Easter, mid May-Oct, daily except Fridays 1400–1700, *Hill of Tarvit Mansion House* (NTS), April, Sat., Sun. only, Easter, May.-Oct. daily, 1330–1730. Tea-room 1330–1645. *Gardens,* all year, 0930–sunset. *Scotstarvit Tower* as for Hill of Tarvit Mansion, who issue keys to visitors. *Griselda Hill Pottery,* Mon.-Fri. 0900–1630, Sat., Sun. 1400–1700.

Tourist office: Cupar. Tel: (01334) 652874.

Ceres – weighhouse.

Cross to the High Street. Next to Brands Inn is the Fife Folk Museum, in the 1673 Weigh House (note the plaque over the door) which once served as the village tolbooth, even to having a dungeon. On the wall outside are the jougs: an iron collar for holding wrongdoers. The award-winning museum has a vast collection relating to life in past centuries and the shop in the annexe across the road has local interest booklets for sale.

Beyond the museum walk down the lane and over the cobbled 17th-century Bishop's Bridge. This leads to a car park, toilets and the Bow Butts where the 700-year-old free games (incorporating the Ceres Derby) are still held on the last Saturday in June. There's a monument to the Battle of Bannockburn, in which Ceres bowmen took part.

At the south end of the car park Woodburn Road leads off left of the row of cottages, a quiet lane, historically called the Waterless Road, being the road from St Andrews to Pettycur and the Edinburgh ferry. Archbishop Sharp came this way shortly before his assassination by Covenanters on Magus Moor in May 1679. As you climb through the rich fields, Hill of Tarvit rises clear. The monument on top is for Queen Victoria's diamond jubilee. Shortly after a row of cottages, you reach Denhead Farm. Turn right round the walled-off house to follow the drive out to a tarred road. This is a private drive, not a right-of-way, and is most attractive. Cross the tarred road for a path continuation. When this comes out on a track turn left, passing a derelict lectern doocot. Turn right at the next junction and down to a ford/arched packhorse bridge then up again to reach a tarred road. Turn right and a few minutes' walk leads to the east drive of Hill of Tarvit Mansion. Enter here (cars are directed to the west drive) and up to the house, skirting a big yew hedge.

The house is an NTS showpiece, notable for its furniture, the kitchen, Chinese porcelain, paintings

(British/Dutch) – and the gardens. There is also a welcome cafe with home baking. From the top of the walled garden, imposing gates give access through woods to reach the open hillside above. Angle up left (steep grass) to the top of the actual Hill of Tarvit, another of Fife's panoramic viewpoints.

The mansion house was built by Sir William Bruce to supersede Scotstarvit Tower. Originally called Wemysshall when owned by the Wemyss family, they sold it to the Dundee jute magnate, Sharp, whose daughter bequeathed it to the NTS, along with its treasures. Sir Robert Lorimer reshaped the house and laid out the gardens in 1906.

Hill of Tarvit mansion and the monument on the hill.

Scotstarvit, a typical laird's tower house, lies 1 km west of the mansion and, if requested, a key can be taken from reception to visit it. Originally built in the 1500s it was bought in 1611 by Sir John Scott, the cartographer (he employed Timothy Pont to map Scotland) and literary enthusiast (his brother-in-law was Drummond of Hawthornden). He married three times and had 19 children.

Leave the mansion by the east drive again, turning left then, after 500 m, left again up the drive to Whitehill. Pass the house and skirt right of the farm, pass the cottages and edge a conifer stand. Big views open out again, right up to the Angus coast. Go through the right hand of two gates for a path down the field edge and rejoin the Moor Road – and 'he tae Cupar mun tae Cupar'.

Once in the town, don't turn right at the station but walk on to a T-junction and turn right into the Crossgate. Pass the post office, spired library building (local booklets on sale) and at the end stands the Mercat Cross and the clock tower of the old town hall. Left lies the Bonnygate, with plenty more shops, cafes, etc., right is St Catherine's Street with the old Corn Exchange building which leads back to the starting point.

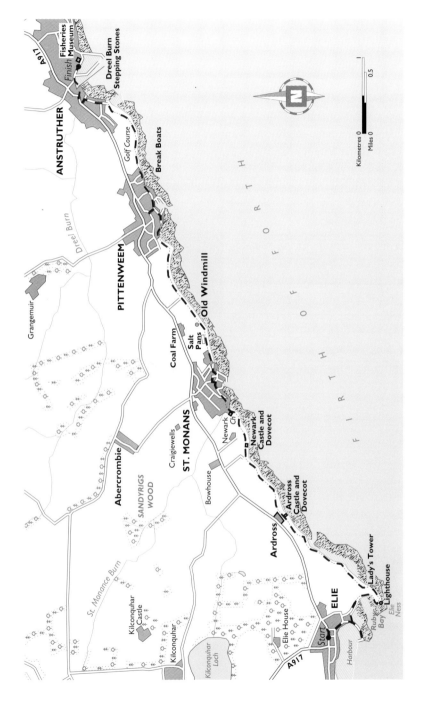

ANSTRUTHER

Fisheries Finish Museum

Dreel Burn Stepping Stones

A917

Golf Course

Break Boats

Dreel Burn

PITTENWEEM

Grangemuir

Old Windmill

Coal Farm

Salt Pans

Abercrombie

SANDYRIGS WOOD

Craigiewells

ST. MONANS

Newark Ch

Newark Castle and Dovecot

Bowhouse

F I R T H   O F   F O R T H

Ardross Castle and Dovecot

St. Monance Burn

Kilconquhar Castle

Kilconquhar

Ardross

Lady's Tower

Elie House

ELIE

Ruby Bay

Elie Lighthouse

Elie Ness

Start

A917

Harbour

Kilconquhar Loch

N

Kilometres 0
Miles 0
0.5

EAST NEUK VILLAGES
(Elie - St Monans - Pittenweem - Anstruther)

This day gives a constant change from town to coast to town so there is plenty to see and do and chances for refreshment *en route*. Work down to the sea at Elie and walk out first to the tidal harbour and the Granary building before heading off along the coast, passing a car parking area overlooking Ruby Bay (the rubies are only garnets!). The coastal path is signposted. Elie Point's lighthouse buildings were built by David Stevenson in 1908.

The next tower was built c.1760 as a summer house for Lady Janet Anstruther, as was the grotto in the back of the bay below. Here her ladyship came to swim – and a bell man was sent round the streets of Elie to warn the plebs to keep away. Dunes lead on towards Ardross, a castle in very ruinous state, but inland, by the farm, is a fine example of a lectern-style doocot.

The path struggles along through a geological chaos to reach Newark Castle, a weathered ruin (with an earlier beehive doocot) which was the home of David Leslie, the general who defeated Montrose at Philiphaugh, near Selkirk, in 1645. Take paths to pass below the church, which is then worth a visit. It is a well-known seamark and has been tastefully restored. Note an old ship model hanging up inside. Cross the burn and go right of the house opposite and climb to a

INFORMATION

Distance: 9 km (5½ miles).

Start and finish: Elie and Anstruther are linked by coastal bus services so returning to the start is not difficult, nor is finding a parking place in central Elie.

Terrain: Coastal paths and towns. Stout shoes advised.

Toilets: In the towns.

Refreshments: In the towns.

Opening hours: *Scottish Fisheries Museum,* Anstruther, Apr.-Oct. 1000–1730, Sun. 1100–1700, Nov.- Mar. 1000–1630, Sun. 1400–1630.

Tourist Office: Anstruther. Tel: (01333) 311073.

Elie.

St. Monans.

Pittenweem – Water Wynd.

road, then turn down again to St Monans (St Monance) West Shore and harbour.

St. Monans is a delightful place and it is worth walking round some of the old streets facing the harbour to see the typical houses, many restored under the NTS Little Houses scheme. When heading east again do not go along East Shore (which looks the route) but up the way and then right, along Rose Street.

Pass an abandoned swimming pool and an area where salt was once produced. Coal Farm reminds of that era. It took 32 tons of water and 16 of coal to make one ton of salt. The windmill is the old pump for these workings. (Key from newsagents in Forth Street.)

Some open country follows, then you climb to a children's play area and, suddenly, there is a view of Pittenweem's West Shore houses rimming the bay. Descend and go along beside them. Pittenweem is now the main active fishing port on this coast with the market and other facilities. At the end of the harbour lie the much-photographed houses of *The Gyles*.

The High Street is high and you can reach it by Cove Wynd which should be Cave Wynd for St Fillan's Cave, which is passed. (St. Fillan was a 7th century missionary.) The key for the cave is held at the Gingerbread House (cafe/pottery) on the High Street. The cave – with its well – is still used as a chapel and the underground stairs led up to an abbey at one time. The Kelly Lodging on the High Street is a 16th century laird's town house and the parish church tower is the old tolbooth. Pittenweem explored, yourselves refreshed, push on for Anstruther.

From The Gyles head up Abbey Wall Road and take an opening to pass a play area on the cliff top. The striated reefs below are, ominously, the Break Boats. The cliff is rather breaking too but follow a path down to the shore again and the edge of Anstruther golf course. Follow along by the shore to Billow Ness and a last bay where fences protect walkers from sliced drives. The battlemented tower on the golf course is the war memorial.

Follow the curve of Shore Road, then right as far as the school, then left up to the main road. Turn right onto the High Street, passing the Dreel Tavern. The road does a sharp bend left to bridge the Dreel Burn but, after noticing the Buckie (Shell) House on the corner, cut down a small road on the seaward side, to

Anstruther. When there, see if you can spot what has changed in the view.

cross the Dreel Burn by stepping stones – if the tide is out. (At full tide you will have to return to the main road.) Plaques here and elsewhere show connections with the clipper ship era, locals owning or captaining some famous names like *Ariel, Taeping, Min* and *Lahloo*.

Anstruther Harbour is the largest of these ports but no longer used by the fishing fleet. Facing the harbour at the far end is the famous Scottish Fisheries Museum, the best of its kind anywhere. There are cafes in it, and in the street behind ('The Tea Clipper'). Anstruther is locally pronounced more like 'Enster', so don't be confused if you hear this.

Allow a good hour for the museum – and refreshments – before heading up to the main road for a bus back to Elie. Information from St Andrews Bus Station, tel: (01334) 74238, or ask locally (Tourist Information Office in the Fisheries Museum), or use a taxi.

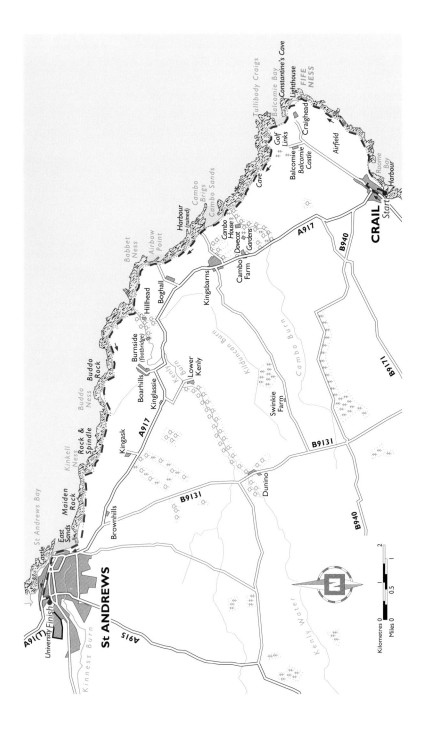

CRAIL TO ST ANDREWS COAST WALK

This is the longest, hardest, loneliest walk in the book, one for the fit and experienced. Years ago some bridges, signposts and markers were put in but erosion and vandalism have played havoc with these and, while restoration work is likely, it cannot be assumed yet. In reality you can hardly go astray: the sea is always on the right! One river has to be paddled (or a long detour made) and there is nowhere offering refreshments. Don't drink stream water.

Crail is a showpiece of vernacular architecture, its tiny harbour a gem, the tolbooth showing Dutch influences (local museum beside it); from there head east along the Marketgate. At the far end is the parish church with a unique collection of early mural monuments, and in season the graveyard and woods are a mass of snowdrops. The stone at the entrance gates was thrown at the church, from the Isle of May, by the Devil. He missed. The stone split, one part falling here, the other on the coast at Balcomie. (These are erratics, left by glaciers, often called 'blue stones' and given superstitious origins.)

Turn seawards along Kirk Wynd and down by a footpath past the 'pepperpot' doocot which is painted white as a seamark. Turn left round Roome Bay, pass the eroded red cliff beyond and join the road along to – and through – the Sauchope Links Caravan Park. Keep to shore-level paths till nearing the knobbly 'castle' rock and then swop to walking on the raised beach level, with the site of a wartime aerodrome on the left. Drop down later and continue (brambles, whin and thorn now the only plants surviving the sea winds) along to the coastguard station and lighthouse marking Fife Ness. The light replaces the North Carr lightship which was originally positioned off the point. Look out for gannets, the Concorde of seabirds, many thousands of which nest over on the Bass Rock.

INFORMATION

Distance: 19 km (12 miles).

Start and finish: Crail or St Andrews but the best choice is to drive to St Andrews (street, car park parking) then bus or taxi to Crail to start the walk. Bus station (office hours information) has timetables on display.

Terrain: Rough coastal walking: proper walking boots and equipment essential.

Toilets: St Andrews, Crail, Cambo Sands car park.

Refreshments: Only at start and finish. Even in winter carry adequate food and drink, suitable for a long, hard walk.

Opening hours: *Crail Museum*, June-mid Sept, weekdays 10.00–12.30, 14.30–17.00. Sundays 14.30–17.00. *Cambo Gardens*, 10.00–16.00.

Tourist office: St Andrews, Tel: (01334) 472021.

Crail Harbour.

Follow a small tarred road to the trim Balcomie Golf Course (founded in 1786), whose seaward edge is then followed. Gritty track leads past Constantine's Cave, where this king was killed by the Danes, c.874. Sandy Balcomie Bay has the old RNLI lifeboat building on it and, at the far end, the second of the blue stones mentioned earlier.

Crags run down into the sea and unhelpful fences reach the tide level so you may be forced to divert upwards if unable to pass between tide levels. Once this obstacle course section is passed easy walking leads to the Cambo Glen, its burn having a substantial bridge.

Cambo was a country park, and a diversion here is pleasant, streamside paths (snowdrops in winter) leading up to the 'Secret Garden', a view of Cambo House and a fine doocot. Back at the shore again take the path on to the warm-coloured Cambo Sands (swimming dangerous). There's a car park/picnic area, the only such public approach to the coast all day, coming down from Kingsbarns.

Very rural walking follows. Beyond Babbet Ness the coast becomes a ragged chaos, this jagged rock being a sandstone laid down in rivers, as against the smoother sandstone from lake sediments. Partridges may go shooting off to make pedestrians jump. Farther on, at an area of sea walls, there are fossil imprints of prehistoric roots. The decaying building is an old salmon fishing bothy.

Keep along the track outside the walled field to reach the Kenley Water – which has to be crossed. At full ebb it is easy but, if paddling is required, choose the wide gravely bit (facing the slip) rather than the rocks more upstream – they can be disastrously slippery. The nearest footbridge is up at Burnside Farm, a tedious diversion. A rock ramp on the far side marks the onward route.

Beyond is a scrubby area (loved of passerines) which can be confusing as it's riddled with paths, so don't get led off inland. Keep to the coast. A boggy area is backed by a secretive lochan and then comes a field with a high wall, which leaves little room to pass if the tide is in, and an uncomfy shore surface if it is out! Things become easier beyond and there's a first glimpse of St Andrews.

Sandstone country (red and greys) leads on to the Buddo Rock 'castle', which has a natural arch and a central passage which allows a scramble to the top. Angle up to follow the top of the bank (wall) to the trig point of Buddo Ness then down again to traverse the wildest section of all, a switchback path through thick scrub, which leads to Kittock's Den, another glen running up inland. Except at high tide you walk on the shore, cross a wall and, beyond, go back to another volcanic area. Traverse a bluff where the ground is giving way; take care. A thorny path leads onto easier ground and along to the Rock and Spindle, a phallic-like pinnacle.

Beyond is Kinkell Ness and soon, or later at some steps, head up onto the heights as the shore becomes almost impossible to fight along. St Andrews is suddenly near and you come on the sprawl of a caravan park, which is skirted before dropping to the East Sands. (Below, another sandstone Pinnacle is the Maiden Rock.)

The East Sands lead on by a Leisure Centre, coastguard building, the Gatty Laboratory and a Putting Green to reach the harbour which is crossed by a retractable bridge. St Rule's Tower and cathedral ruins above show the route into town. You will come onto North Street which is followed to the far end (roundabout) where, turning left, you come to the Bus Station. St Andrews is so rich in sites/sights that Walk 18 is entirely given to its exploration. A busy tourist town, it has a large selection of accommodation, restaurants, etc. Try Baburs (89 South Street) as an exotic ending worthy of this energetic coast walk.

St. Andrews Harbour.

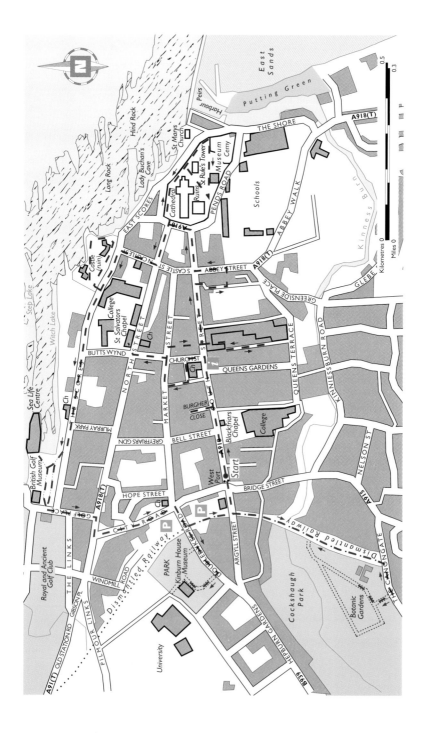

EXPLORING ST ANDREWS

T his urban day is quite tiring if most of the sites are visited. Ask for the West Port, the old walled gateway to the town.

Walk along attractive South Street to the isolated ruin of Blackfriars Chapel with the lawns of Madras College behind. A notice describes the school, founded by a missionary to India, its 'suave Jacobean manor' appearance hiding early austerity. (One rector wrote of donning two sets of woollen underwear in October and keeping them on till April.)

Cross the street to go into the quiet Burgher Close and, in Hansa House, visit the studio of Jurek Pütter who has spent a lifetime researching medieval St Andrews. Out again in South Street turn left and, passing the Town Kirk, cross diagonally to the Town Hall (clock over street) to find the Tourist Information Office.

The Town Church (Holy Trinity) is historic but may not be open. The huge mural tomb to Archbishop Beaton, murdered in the castle in 1546, is a notable feature. John Knox preached an epoch-making sermon here in 1559. The chancel floor is of Iona marble, the roof is of Caithness slabs.

Go through Logie's Lane behind the church and right onto Market Street, once the heart of the town with cross and tolbooth, and still observing the ancient Lammas Fair. It was also the site of executions: of the hapless Chastelard (see Walk 8) and of Paul Craw, a Bohemian reformer burned in 1433. Cross over and along College Street to reach North Street, dominated by St Salvator's Tower, a splendid medieval survival. On the pavement by the *pend* (passage) the initials P.H. are for Patrick Hamilton, the first Reformation martyr, burned here in 1528. Through the passage, right, is the church entrance. The university sprawls all along North Street. Only Oxford and Cambridge are older foundations.

INFORMATION

Distance: 4 km (2½ miles).

Start and finish: Anywhere in the town but avoid expensive parking streets. Car parks are signposted. It is worth picking up a town map from the Tourist Information Office, 78 South Street KY16 9JX, Tel: (01334) 72021, before commencing the walk.

Terrain: Streets and roads, ordinary shoes adequate. St Andrews can be windy and cold, so be clothed suitably.

Toilets and refreshments: Available throughout the town and at the museums, etc.

Opening hours: *Cathedral Museum,* all year, Mon.-Sat., 0930–1800, Sun., 1400–1800; *Preservation Trust Museum,* daily mid June-mid Sept. 1400–1630; *Sea Life Centre,* all year, daily 1000–1800; *British Golf Museum,* open daily, May-Oct., 1000–1730, other seasons closed Tues./Wed.; *St Andrews' Museum,* Apr.-Sept., daily 1100–1800, Oct.-Mar., Mon.-Fri., 1100–1600, Sat., Sun., 1400–1700; *Botanical Gardens,* open 7 days a week, 1000–1900, May-Sept., rest of year Mon.-Fri., 1000–1600; *Glasshouses,* weekdays only, 1400–1600.

Return to Market Street, cross leftwards, through Crails Lane and over Smith Street to the arched entrance of St Mary's College. The quadrangle has a holm oak planted in 1728 and a centuries older thorn. Explore the gardens and, out, turn right along South Street.

Turn into South Court Pend (Byre Theatre sign) where there's a plaque to J D Forbes, the alpinist who showed a surprised world that glaciers were moving rivers of ice. Byre is *cowshed* and the tiny original theatre was created in a byre in 1933. The present theatre building is 1970. Once through, turn left and regain South Street. Nearing its end the cathedral ruins are framed between the (leaning) Roundel and Queen Mary's House. The Pends gateway led to an old priory and down it, right, is St. Leonard's School, a building which began as a 12th century hospice. The chapel can be visited.

Facing the Pends is Deans Court, once the Archdeacon's residence. The cobbled symbol marks where 80-year-old Walter Myln was burned. Go through the gate into the cathedral grounds. Once the second largest in Britain, it is now a 'rent skeleton' (Ruskin) and the main interests are St Rule's Tower (small charge for climbing 33 m to the viewpoint on top), the museum, and wandering round to look at the monumental stones, which range from full-rigged ships to golf champions, including young Tom Morris who won four Opens in a row before dying, aged only 24. The cathedral suffered at the Reformation and its stonework was pillaged for town buildings. The museum has a collection of early stones, several 'green man' carvings, a skeleton stabbing a victim in the back and, beautiful beyond description, an old sarcophagus panel alive with the animals of a hunting scene.

Leaving the grounds, walk along North Street. The Preservation Trust's Museum is on the left. Its mock-up of a Victorian shop is entertaining. Turn right into North Castle Street and ahead looms the castle itself. The G.W. initials outside are of the reforming

St Andrews – the view from the top of St Rule's Tower.

preacher George Wishart, burnt here in 1546 while Cardinal Beaton watched from a castle window – out of which his body was soon to hang. There isn't much castle but unusual are the mine and counter-mine (which you can explore) and the notorious bottle dungeon (which is not recommended!). The 1546 mine was cut through solid rock to place charges under the walls to blow them up but the defenders burrowed a counter-mine and intercepted the attempt.

Turn right on leaving the castle to walk along The Scores to reach the obelisk of the Martyrs Monument. Below it are two modern attractions. The Sea Life Centre is a wonderful window into the sea world (and the balcony cafe over the sea may be very welcome) while the British Golf Museum uses every high-tech device to captivate anyone even slightly interested in golf. The building opposite is the Royal and Ancient clubhouse with the Old Course beyond – the very home of the sport.

Royal & Ancient clubhouse.

The West Port can be regained from the museum by walking along Golf Place, turning right, then first left along past the Bus Station. If legs and brain can stand any more, off Doubledykes Road (right at the roundabout beyond the Bus Station) is a new town Museum which also uses modern techniques so you can see, hear and even *smell* the past. From the West Port if you head off along Argyll Street (the opposite direction to the morning's start) you soon see a sign indicating the Botanical Garden. This is a ten minute walk away and besides the attractive layout has a cacti collection (250 species) and an orchid house.

You may well have to be somewhat selective with the above. If less interested in the historical sites, the museums are fun, and from the Golf Museum you are close to the huge West Sands which are a favourite walk of 'town and gown'. The opening shot of the film *Chariots of Fire* showed the athletes running along the tideline here. The famous golf courses lie inland from the sands.

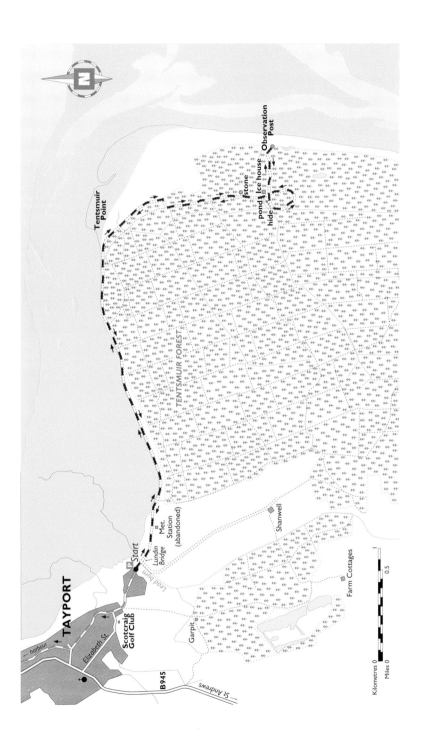

TAYPORT

harbour

Scotcraig
Golf Club

Elizabeth St

B945

St Andrews

Garpit

Leod burn

Start

Lundin
Bridge

Met.
Station
(abandoned)

TENTSMUIR FOREST

Tentsmuir
Point

stone

pond
hide

Ice house

Observation
Post

Shanwell

Farm Cottages

Kilometres 0

Miles 0

0.5

TAYPORT AND TENTSMUIR FOREST

Tentsmuir is a huge forest, started in 1922 on what where basically salt marshes at the north east corner of Fife. Like most forests the main interests and wildlife are found round the edges and open areas so our walk keeps to the coast in both directions. (Circuits through the forest interior are a dull option, navigation can be difficult and felling interrupts progress.) Some of the area is a National Nature Reserve, of which more later. Take a picnic and plenty of liquid, and go early (and quietly) to maximise the chance of seeing wildlife. Most of the trees you'll see are Scots pine.

The easiest approach to the start is from the St Andrews road. On entering Tayport the striking white church of Our Lady Star of the Sea is seen, left, and then, right, Elizabeth Street is indicated with a sign for Golf Club. Head down to pass the Scotscraig Golf Club and at a T-junction, turn right, Shanwell Road, and follow this to its end. Park on the edge of the grassy area looking onto the sea and Tentsmuir Point.

Head off round the coast from Lundin Bridge. The tiny Lead Burn hardly seems sufficient to have powered mills last century. Keep outside the old Meteorological Station, then follow the track in to the corner of the forest. The shore is barred by concrete blocks laid as anti-tank defences in the 1939–45 war when invasion was a possibility. There's a confusing number of tracks but turn left (almost at once) at an open space to walk along just outside the permanent forest on a good track (trees on the left are self-seeded). Gaps to the left allow views over to the Angus coast. The sands at low tide can be 1 km wide.

There's one 100 m stretch where the sea is winning over the land and the pines are tumbling from erosion. The path has gone so cut along through the trees or walk the sand, depending on the tide. (The concrete

INFORMATION

Distance: 13 km (8 miles).

Start and finish: Lundin Bridge, south of Tayport, reached via the golf course approach road.

Terrain: Sandy paths and tracks in forest. No special footwear needed.

Toilets: Tayport.

Refreshments: Tayport.

The drive to Kinshaldy – Tentsmuir Forest.

blocks have been tossed-about by the storms too.) The next landmark is a gas pipeline marker, then a drain, and finally the point is reached (lots of broom).

The path turns, along by a fence, but east the land is still building up steadily as sediments are brought down by the River Tay. (Some of the grains you walk on originated on Schiehallion or Ben Alder!) Scientists are taking the opportunity to study this natural reclamation and all the land east of the fence is a reserve (SNH). The concrete blocks are beside you – but seemingly miles from the sea!

Some larger areas have been fenced-off and are marked by a gate/stile with a reserve board. Turn right (inland) and then left at the T-junction with a major forestry track. It passes through some other fenced-off sections, the netting right onto the ground to try and keep out rabbits. Less than 100 m past this open space (left) keep an eye open for a stone like an elongated milestone. The inscription shows it to be a 1794 boundary stone between two salmon fisheries – and is lined on Norman's Law. It once stood at the edge of the shore!

There's a fork with one branch of track turning inland, but keep straight on and in a few minutes reach your destination so to speak, the large ice house on the left. Ice houses were half-buried and enclosed like this in order to keep the ice from melting in the days before refrigeration was available to fishermen. Ice could even be carted from the Highlands. This must have been a remote and desolate spot in those days.

The old ice house in Tentsmuir Forest.

Just beyond the ice house area a small footpath goes off right and if followed (marker posts) leads to an observation hide looking onto a pond. Tentsmuir has a very diverse birdlife. You may see heron, moorhen and gulls here, and already have noted wren, green and great spotted woodpeckers, sparrow hawk, curlew, warblers, etc. And colourful snails, rabbits galore, foxes, roe deer and red squirrels. Go quietly and you should see most of these. Bat boxes have encouraged three species to live in the forest.

The pond in the heart of Tentsmuir Forest.

Nearing the hide the path forks. Follow the other fork thereafter. It takes the line of a beech avenue then swings onto a forestry track. Turn left and it brings you back to the ice house area. On the seaward side there's a track going to a gate/stile with a reserve sign. Take this and walk out to the dunes at an old wartime look-out post. The concrete blocks run along just in front of it, quite invisible as they are under the dunes.

This is the place to picnic. Seals haul out on the distant sands. There are dangers so do not swim or even go out any distance from the tideline. If dunes fascinate, a separate visit is recommended to Kinshaldy at the south end of Tentsmuir Forest. Well signposted from the Tayport - St Andrews road, there are marked trails, picnic and barbecue areas, toilets, etc. You now wander back to Tayport.

Leaving where you parked, drive straight on instead of turning left to the golf club. This becomes Nelson Street. At its top end take the second right which leads along to the attractive harbour area (Bell Rock Tavern a landmark) where there are toilets by the car park. Tayport (Ferry-Port-on-Craig) was for long the great crossing point for the Tay. Broughty Ferry, on the other side, is still marked by a castle (museum). The Tay Bridge ended the prosperity of ferry days. Broad Street/Castle Street leads up as the main shopping street and, turning right at the end, heads you along to the Tay Bridge and wherever. A rude old saying has 'Out of the world and into Fife'; for Tentsmuir it is rather apt.

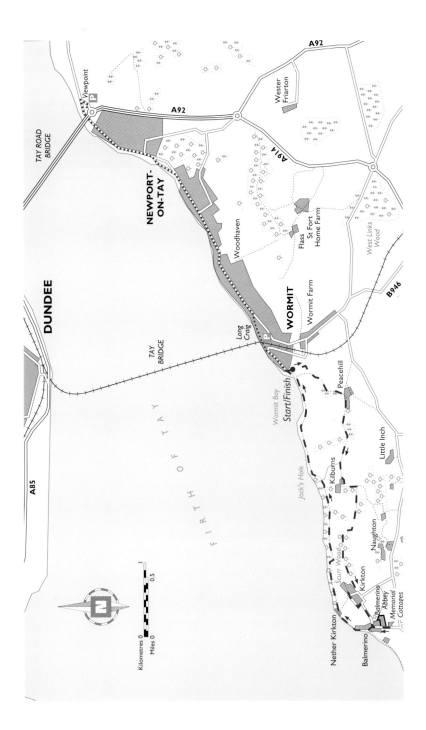

BALMERINO AND THE TAY ESTUARY

Pause when driving in to Wormit to enjoy a close view of the Tay Bridge. (There is a parking area near the signal box; once Wormit Station.

The bridge of course is the second. The first bridge collapsed in a storm in December 1879 taking a train with it at the cost of 75 lives. The stumps of its pillars can be seen just east of the present 85 span, 3½ km sweep.

Wormit grew up as a dormitory town for Dundee following the establishment of the railway bridge. It was the first town to be lit by electricity and has plenty of assured Victorian buildings.

Driving on into Wormit note the 'extravaganza in black and white' shop-on-the-corner building (right) then, by the blacksmith's, turn left onto Bay Road and follow it to its end. Parking by the shore.

Take the path up the edge of the field by the last house, passing above a pond at the foot of Wormit Den and then climbing steeply to swing right along the field edge, then left, to a gate, beyond which the track (the old Woodhaven - Edinburgh road) is clearer up to the farm of Peacehill. Peacehill itself is a name of black humour for a gallows once stood there.

This corner of Fife is intensely farmed (no set-aside here!) and grows vegetables like early potatoes, lettuce and broccoli beside various cereals and raising beef cattle and, here you see vast sheds of poultry. Keep right of the buildings and turn right along a track at the farmhouse itself. The track passes several cottages (one with pigeon lofts) and runs out to Kilburns Farm. There's a concrete/grass track which bypasses Kilburns on the left to reach the farm's drive up by the tree shelter belt.

At the top turn right (the track swings left) but unlike Lot, do look back – a grand view. This track runs

On the Balmerino - Wormit coastal path.

along by Hay's Hill plantation and then you follow on in the same line towards Scurr Hill Wood. Angle down outside it till the ground falls away steeply and only then cross the fence onto a footpath through the wood. Exit by a gate and turn up to continue first, outside the plantation, then inside it (a real tunnel section) and eventually down outside the back gardens of some Kirkton houses.

You come up to an obvious garden gate leading into someone's garden but, no panic, the right-of-way goes through this gate, then sharp right, down by the house, and out a similar gate onto a minor road! Turn left, ahead at a crossroads and right at a junction to pass the local cemetery and war memorial. A stile at the memorial gives access to the cemetery. Second stone in, second row, has some interesting spellings and upside down letters, another stone has an unusual crook symbol (shepherd). The church is at Bottomcraig, not Kirkton, and Balmerino (pronounced Bamérnie) is only the hamlet you pass at the foot of the hill. Turn left and look out for the sign for Balmerino Abbey.

The Abbey (NTS) dates back to 1226 but was burnt by the English in a 1547 Commando raid and pillaged after the Reformation. Vaulted sacristy and chapter house survive but the most unusual object is a Spanish chestnut tree nearly 500 years old. The stone for the Abbey was brought by sledge and water from Strathkinness, near St Andrews.

A short distance on you come to an unusually formal square, which is a war memorial from 1948. The story is told in the Doric-pillared portico. The row of statues depict the four seasons. The cottages are lived in: a very practical memorial.

Head back seawards again, onto the track marked 'No through road', which reaches the shore by old mill buildings. The white cottages farther on are converted fishing bothies and as Nether Kirkton is reached there's a large, round erratic on the beach, called Samson's Stone. 'Footpath runs between house and shore' you are told but, past the buildings, it slants up and along (white marker posts) to follow the top of the wooded cliffs, at first just inside the trees, then simply along the field edges with Scurr Hill above. Rabbits galore. There's one deep-cut den to cross (below Kilburns) and then the view really opens out. Use an obvious gap in the fence and follow cattle trods along. Whin gives way to dog-rose and elder as the season advances and rape fields can be scented from miles off.

Nearing the bay, be sure to keep to the fence as the path dives down a buckthorn tunnel to reach sea level. The view up the Tay is pleasant and the rail bridge sweeping off to Dundee remains impressive.

Back at the start, you should not miss driving east towards the 1966 road bridge, both for the views and to enjoy the swanking villa architecture of the bold Victorian jute barons who built their houses here – well aside from their smoky factories in Dundee. Woodhaven was an early ferry point but lost out to Newport. Dr Johnson crossed from there and, in his journal, complained at the fare.

Cow's Hole and Pluck the Crow Point lead to Newport-on-Tay, with no break in the buildings but the town centre is marked by several shops and, off Cupar Road, a cluster of churches. Don't miss the 1882 canopied drinking fountain above the sea: wrought iron at its most elaborate. Thomas Telford was the engineer of Newport's harbour and ferry pier. Did you notice the 'Pettycur' milestone outside the ferry building? It gives the distance to Newport as 0.

Drinking Fountain – Newport.

Drive under the road bridge then turn first right (Tay Bridge sign) and into a car park. The 2 km incline gives the bridge an exaggerated perspective as seen from the 'aeroplane-wing' monument. The bridge leads right into the city of Dundee but if you think 'Great, no toll!' going in, be prepared for a double toll on leaving. *Discovery Point* and other features have given the town a new tourist interest.

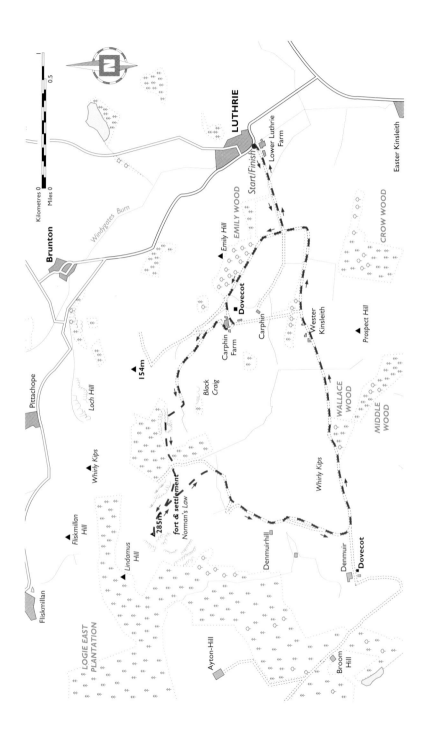

NORMAN'S LAW

Park tidily in the village and set off past Lower Luthrie farm (footpath sign for Ayton) on a long straight road. At the end of this straight, turn right to go up past a house and along outside Emily Wood. The path then dips and rises to Carphin Farm. In the field, left, is an unusually slender lectern doocot. Swing on up through Carphin and, after a levelling-off, take a left fork track to climb steeply again (gate). Rolling green fields, hotching with rabbits, and with rooks calling from the Scots pines, lie over to the right.

The track keeps swinging left but, on nearing an obvious col, break off to climb the hill on the right. There is a sudden feeling of height for underfoot are blaeberry, heather, tormentil, bedstraw, milkwort . . . and the summit is bare rock and the view wide with the first sight of the Sidlaws, the Tay estuary and Dundee. Norman's Law looms ahead.

Head towards it but slip down the left flank of the hill to avoid some cragginess and on along outside the fenced area to a small gate which gives access to the final slopes. There's another subsidiary bump to go

INFORMATION

Distance: 8 km
(5 miles).

Start and finish:
Luthrie village, off
A914 (Kirkcaldy - Tay
Bridge road).

Terrain: Mostly
roads, tracks and paths
but rough, open
ground on hill itself.
Strong footwear
advised.

**Toilets and
refreshments:**
Cupar, 5 miles;
Newburgh, 8 miles.

Brunton and Green Crag
from the road over East of
Norman's Law.

The thin doocot at Carphin Farm on the way up Norman's Law.

over before Norman's Law will be gained. There are iron age encampments and a fort on top, the defences of the latter very clear. The situation is fantastic. 'The view's better than from the top of Ben Nevis,' I've heard proclaimed – and wouldn't argue.

There's Highland hills looming from Ben Lomond to Lochnagar (Lawers, Schiehallion, Farragon, Beinn a' Ghlo, etc.) and a view indicator will help pick them out. There's also a trig point and a large cairn. The view up-river is exquisite, over Fife, varied and in early summer, patched with the bright yellow of rape fields and the dark green of conifer plantations. This is a summit for a late evening visit to catch the sun dipping westwards.

The names around here are intriguing. On the Tay we'd expect a Dog Bank but what of Eppie's Taes Bank, or Sure As Death Bank or Peesweep Bank? There are cheery land names: Robin's Brae, Cherrybank, Sunny Den and Blinkbonny; awful names: Dark Law, Foodie, Gallow Hill; amusing names: Lindifferon, Glenduckie; and try getting your tongue round Mountquhanie or Cunnoquhie.

Start off down by the way ascended, as far as the first dip, then peel off right, heading south-east under all the rockiness of the minor bump to meet a fence and a gate with a minor track. Follow this rutted, green way along and then twisting down to a gate into a field whose edge is followed. The slopes of whin, right, can be heavily scented in May/June.

The track is clearer after the gate out of the other end of the field and angles along on the flank of Whirly Kips. Denmuir, below, is a large farm and has a big lectern doocot visible. The track finally swings left, bringing the often-seen pencil of the Hopetoun Monument dead ahead, goes through a gate with an interesting catch, then curls off down to Denmuir. You however, turn left on a less-used track but one which long predated the busy A-roads of today down on what, in those days, was the boggy low ground. Robert Baillie of Luthrie created many of the 18th century roads hereabouts.

Denmuir and its doocot from below Normans Law.

You simply follow this old route from Ayton and Denmuir over to Luthrie, at first up a near-tunnel of thorn, elder, wild roses and gean, then with fields on the left allowing good views up to Norman's Law again, then as a lane to Wester Kinsleith and, finally, a firmer avenue down to the start, thus completing a memorable round. But do keep Norman's Law for a clear day. Cattle and sheep are grazed in many places so dogs are best left behind, and young children could find it too hard.

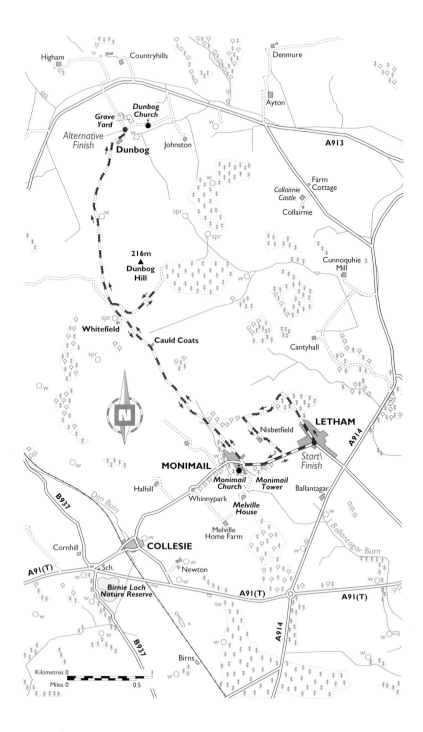

Higham

Countryhills

Denmure

Ayton

Grave Yard

Dunbog Church

Alternative Finish

Dunbog

Johnston

A913

Farm Cottage

Collairnie Castle

Collairnie

216m ▲ **Dunbog Hill**

Cunnoquhie Mill

spr

spr

w

Whitefield

Cauld Coats

Cantyhall

spr

N

Nisbetfield

LETHAM

A914

MONIMAIL

Start\ Finish

Halhill

Monimail Church

Monimail Tower

Ballantagar

Whinnypark

Melville House

B937

Den Burn

Ballantagar Burn

Melville Home Farm

Cornhill

COLLESIE

A91(T)

Sch.

Newton

A91(T)

Birnie Loch Nature Reserve

A91(T)

A914

B937

Birns

Kilometres 0   1
Miles 0   0.5

LETHAM, MONIMAIL AND DUNBOG

This is another corner of Fife which reflects times past, taking in the hamlets below the hills north of the Howe of Fife yet, seeing the prosperous modern farms, the past is almost inconceivable. Letham at the start of this century had a shoemaker, blacksmith, saddler, cartwright, joiner, tailor, baker, grocer and butcher, there was a malt barn and every cottage had its loom.

Letham is a common east coast name so make sure you find the right one! Turning off the A914, park somewhere on the street of little cottages. At the top, on the west side, is the surviving village pump. Set off up School Brae, but first go to the far end of the row of houses right of the brae and keek round the back to see a charming pepperpot doocot (private garden, so don't go in please). Halfway up School Brae there is a lectern doocot, right, and where the road forks, if you walk left a bit you'll see another. Return and continue on the track up into the hills.

At the end of the trees, fork left round the wood to traverse the south-facing hillside with its fine view. Close all gates. Turn left on the farm track down to Nisbetfield. Monimail church is seen over its roofs and is your next port of call, an elegant whinstone building (1794) with a later tower. Clear windows allow a look

INFORMATION

Distance: 7 km (4½ miles) to Dunbog; 9 km (5½ miles) returning to Letham.

Start and finish: Letham (7 km west of Cupar, on a minor road between the A914 and A91/B937 junction).

Terrain: Good tracks, minor road.

Toilets and refreshments: None, but Cupar and Newburgh not far off.

Letham.

Monimail's rich rural landscape.

inside. Walking on you come to Monimail, a place which seems about to be swallowed by jungle. In summer the scent of ransoms hangs on the air. The churchyard is worth a visit, with war memorial gates and little stories on the stones. There are plenty toffs' mausoleums, including the large Melville aisle of the original church. Another is full of Balfours of Fernie Castle and, *outside* the entrance, a Balfour who was just the local blacksmith – for 50 years.

Turn right from the gates to go through a hidden door to view the Monimail Tower, best known as part of the palace home of Cardinal Beaton, murdered at St Andrews in 1546. His successor as archbishop, Hamilton, was hanged in 1571 and eventually the estate was acquired by the Melvilles who abandoned the tower to build sumptuous Melville House. If you walk down the drive from the immaculate North Lodge you can glimpse the mansion (now a school) then turn back and cross the road to head north into or over the Dunbog Hills. Monimail House is the old manse and you pass the new graveyard. Monimail Tower can be visited and there are efforts being made to try and restore it and other long-abandoned features in the walled garden.

The track follows a long valley. The farmstead of Whitefield which sits on the col at its head has been restored. Shortly before, the ruins on the right are of Cauld Coats, a name appearing on a 1775 map, which shows the way from Money Meal (*sic*) to Dunbog but not the modern road. As the track turns to its end at Whitefield go through a gate and straight on, to have one of those instant views which are unforgettable. You could turn back from here, but it is worthwhile going to the next gate and through it to the good track beyond.

Turn right up this track for a few minutes only, simply to claim a view in the other direction too, probably over oilseed rape fields to the pillar of the Hopetoun Monument and perhaps with skylarks paragliding overhead. Don't go to the end or up Dunbog Hill, but either head back to the car via Monimail and the minor road or, if you have arranged car help, take the track right round and down to Dunbog, which is actually the shorter option.

Descending north there's an unusual view through the Lindores gap to the waters of the River Tay, and Dunbog church occupies a wide east-west vale so is very prominent. It is now a private house. The graveyard's main interest is the remnant of its watch house with the inscription: '1822 Erected for protecting the dead.' If being met at Dunbog, both Newburgh and Cupar are reasonably handy for refreshments.

If returning to Letham drive west, through Monimail again, to see Collessie, another hamlet of character with a dominant church and one row of weavers' cottages restored as a house with a thatched roof. Between Monimail and Collessie the road does a wiggle at a cottage. This belonged to a weaver whom the Earl of Melville tried to remove in order to straighten the road. The cottage – and the wiggle – remain.

Beyond Collessie there's a double junction but if you turn left onto the A91 (Cupar direction) and, almost at once, turn right you will find Birnie Loch Nature Reserve on the left. This was created by the owner of a quarry, after the site had been worked out, and given to the local council in 1992. Already over 50 species of birds have been noted (including osprey!) and when the 3000 new trees grow higher it will be a marvellous place.

Fernie Castle near Letham.

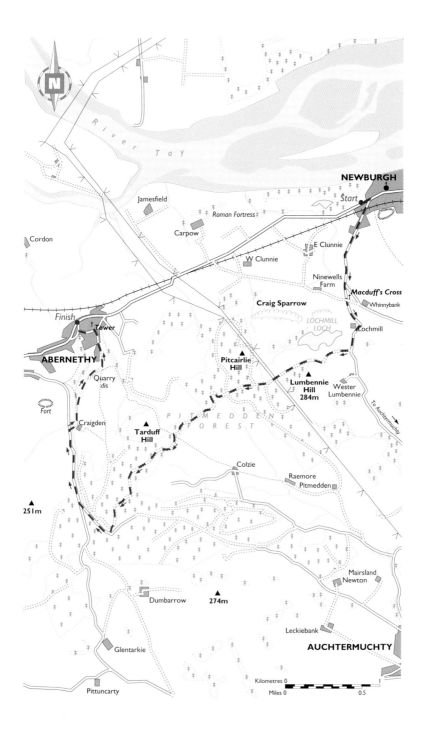

NEWBURGH TO ABERNETHY BY PITMEDDEN FOREST

Newburgh ('new' in 1266!) and Abernethy are both interesting old towns squeezed in between the rump of the Ochils and the big River Tay and both deserve some exploration. Every house along Newburgh's High Street seems to be attractive and the whole is something of an 18th-century showpiece.

The town came into being in connection with Lindores Abbey (1178), the scant red ruins of which lie at the east end of the town. It had many royal connections and bloody history but the records have nice touches, such as the monks having Papal blessing to wear bonnets because of the climate, and their constant feud against adders. They introduced fruit-growing to the area. The closure (and demolition) of the linoleum works has left the town rather lifeless. Newburgh falls down to the river opposite huge reed beds, some of which are used for thatching houses, another near-extinct craft. At the western edge of the town is a large car park, next to the bowling green and war memorial. The walk starts up Woodriffe Road, opposite the war memorial.

The road climbs steeply and opens out a grand vista over the Earn and Tay, a view that had Sir Walter Scott rating it world class. Keep right when the road forks and, at cross tracks (Ninewells Farm entrance) turn left to see the historic MacDuff's Cross, a cup-marked sandstone block, to which many legends are attached.

Continuing, there's a glimpse through to Lindores Loch. The road steepens and twists up by Lochmill Farm with the wall of a dam above it and when it tops the crest there is a track going off right, which is your route. There are views down to Lochmill Reservoir as the track climbs along by Craigdownie and the forest edge of Lumbennie Hill. Power lines cross the route

INFORMATION

Distance: 9 km (6 miles).

Start: Newburgh, on the River Tay, east of Perth. Car park at west end.

Finish: Abernethy. A bus service operates to Newburgh.

Terrain: Road, forest track and good footpath. Have comfortable footwear.

Toilets and refreshments: Newburgh and Abernethy.

Opening hours: *Abernethy Round Tower,* 1000–1700 weekdays, 1200–1700 Sun., key from cafe opposite; *Laing Museum, Newburgh,* Apr.-Sept. Mon.-Fri. 1100–1800, weekends 1400–1700; winter, Wed., Thurs. 1200–1600, Sun. 1400–1700.

The Pictish symbol stone – Abernethy.

Lochmill Loch.

and buzzards may be heard mewing overhead. At a T-junction turn right. Having been felled and replanted there are better views down on the Pitmedden Forest side. Drop down to the evocatively named Seven Gates, a multiple junction.

Head straight on, twisting up and then walking with magnificent views into Fife. Ironically you are in Perthshire here, with what many regard as the finest view of the Lomonds of Fife, while earlier, at Scott's View (MacDuff's Cross), you were in Fife but the view was of Perthshire. It's ironic too that this large area of conifer planting is called Pitmedden Forest when the Pitmedden hollow below you (with its yellow collar of whin) is about the only area not afforested.

The forest closes in for a awhile, then you begin to lose height; on the right there's a greened-over lochan, then you come to the edge of the old forest again, being joined by a track from Newhill and Reedie Hill (red deer farm). Outside our forest is the young Glen Tarkie forest covering Dumbarrow Hill. Descend steadily to Abernethy Glen and join the road from Perth into Fife (Strathmiglo), turning right for 1 km to Craigden. Watch out for speeding cars on this twisting road.

At the entrance to Craigden, turn left, for a footpath down Abernethy Glen. This goes by the Ballo Burn then steps lead up out of the 'den' onto the Witches Road, a traversing path named after a real enough 'coven', judicially burned and their ashes scattered on Abernethy Hill in the 17th century. Looking across Abernethy Glen there is the prominent bump of Castle Hill with an old quarry on its slopes. There's a prehistoric fort on the prow. Preaching How perhaps refers to Covenanting times and Quarrel Knowe indicates where archery once took place. There's history in map names! You come out at Loanhead Quarry, the track becoming Kirk Wynd as it leads down to the square.

Abernethy's landmark feature is its 'Irish' Round Tower, dating to the 9th century and, with Brechin, the only survivor of its type in Scotland. Unlike most Round Towers this one can be climbed. The churchyard is also worth a browse and, on the tower,

Abernethy round tower – one of only two 'Irish' towers left in Scotland.

note the jougs (collar for malefactors) and an old Pictish symbol stone. Earlier the Romans had a ferry at Carpow.

The first church here was AD *c.*460 and the Pictish kings made Abernethy their capital. Only the tower remains to indicate any glorious past. Turn west to get down to the main road, the A913; there is a bus stop opposite the one-time coaching inn (Abernethy Hotel). You pass Tootie House, a reminder of how a herd would blow a horn each day to see folk took their cattle off to the common grazings.

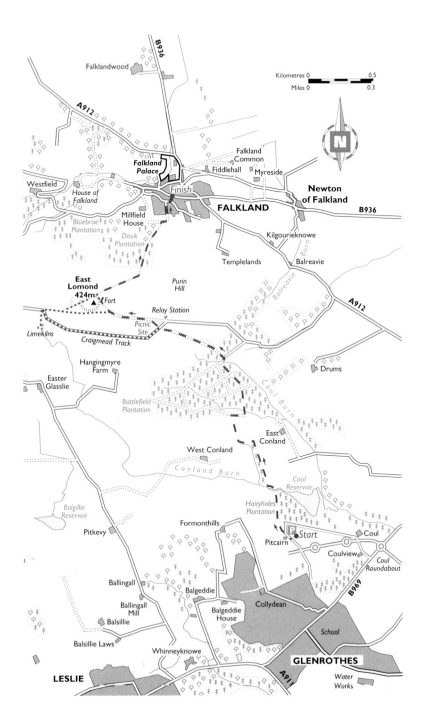

B936

Falklandwood

A912

Kilometres 0   0.5
Miles 0   0.3

N

Falkland
Common

**Falkland
Palace**

Fiddlehall   Myreside

Finish

Westfield

House of
Falkland

**Newton
of Falkland**

FALKLAND   B936

Millfield
House

*Bluebrae
Plantation*

Kilgourieknowe

*Douk
Plantation*

Templelands   Balreavie

**East
Lomond
424m** ▲ *Fort*

*Purin
Hill*

Balreavie Burn

Relay Station

A912

Limekilns

*Picnic
Site*

Craigmead Track

Hangingmyre
Farm

Easter
Glasslie

*Battlefield
Plantation*

Coul Burn

▢ Drums

West Conland

East
Conland

Conland Burn

*Coul
Reservoir*

*Balgillie
Reservoir*

Formonthills

*Hairyholes
Plantation*

Pitkevy

🅿 *Start*

Coul

Pitcairn

Coulview

*Coul
Roundabout*

Ballingall

Balgeddie

Collydean

B969

Ballingall
Mill

Balsillie

Balgeddie
House

Balsillie Laws

Whinneyknowe

School

GLENROTHES

LESLIE

A911

*Water
Works*

EAST LOMOND AND FALKLAND

T his day gives a crossing of the Lomonds *and* a walkabout in historic Falkland. Being a traverse route a second car must be left or a willing driver drop and pick up the walkers. East Lomond (Falkland Hill), at 424 m, is the second highest point in Fife and has a sweeping summit view.

Glenrothes sprawls up the lower slopes of the Lomonds so follow directions carefully to the start, from the main road skirting the east side of the Lomonds (A912, A92).

Turn off on the B969 and at the first roundabout (Coul Roundabout) turn right for Pitcairn. Coming from the west (Leslie) turn left at the first roundabout onto the B969 and the Coul Roundabout is the second roundabout thereafter. Turn left for Pitcairn. (A pseudo stone circle decorates the roundabout).

Go straight on at the next roundabout then, at a second, turn right into an area being developed, where a sign indicates a small road, left, for Pitcairn and leads up to a car park. The cottages are the Ranger Base and the start has a notice board and map.

This is a Public Footpath Agreement route in the Fife Regional Park so tends to follow wood edges in angular regimentation and early views are spoilt by power lines. The first plantation bears the name Hairyholes and there's a dip down and up to cross the Conland Burn. An open field leads up to the larger Battlefield Plantation and the route becomes a forest track for a while. This comes to a T-junction and, turning right, there is a long view along the Howe of Fife. The path then heads left, up the short stretch of forest remaining, to gain the blaeberry/heather moors at a delightful beech clump. Rowans are sprouting as there is little grazing. An old wall is followed up to the car park by the relay station (notice board, picnic tables, toilets).

INFORMATION

Distance: 6 km (4 miles) direct.

Start: Pitcairn Ranger Base near Glenrothes (details below).

Finish: Falkland (big car park signposted)

Terrain: Mostly footpaths or in town but the steep hill calls for boots and fitness.

Toilets: Car park on East Lomond. Falkland.

Refreshments: Wide choice in Falkland.

Opening hours: *Falkland Palace,* Apr.-Oct., 1100–1750, Sun. 1330–1730; *Town Hall,* 1330–1730 only.

Less than 1½ km west along the Craigmead track lies a restored limekiln with interpretive display boards which could be taken in on a half-hour diversion.

From the car park cut through on to the track which runs along by the old wall for a bit (willow scrub) then goes through a gate (warning sign to keep dogs on lead) for the final upthrust of the hill, which is a remnant of a volcanic outpouring. Oddly, a trig point lies off left on a lower shelf; the bare, flat summit only has a view indicator.

Tackling the final steep slope, Falkland appears, down right, with a path leading clearly to the plantation above it. This is your route down and to prevent erosion on the direct descent (which is dangerously steep anyway) please return here to head down at this easier angle, joining the path at the gate through a fence.

The summit view takes in Highlands, Lowlands and the Southern Uplands as well as the sweeping Forth Estuary. The rest of the Lomonds are close and, below, lies the restored limekiln. Beyond Craigmead the central bump among the bumps is Maiden Castle, a prehistoric fort, as indeed is the summit where we stand. (Down on the Falkland side the circling walls are still clear.) Norman's Law (and the Hopetoun Monument) and Largo Law are other walks clearly seen.

Head down as indicated above. The path from the gate wanders down the moor to reach the Douk

One of the historic inscriptions in Falkland.

Plantation. Down this, you are glad to leave the crypt-gloom of conifers into a cheery choir of beechwood. An opening gives a view to Falkland's spires and palace and 'the works'. A steep slope leads down to a driveway; turn right and down into Falkland.

The St John's Works began as a weaving factory but sixty years ago turned to making linoleum. Now the works produce plastic bags. Turn right just past the works and, when the road swings left, you are suddenly in 'old' Falkland. Pause at the first crossroads.

The house on the right has a 1663 marriage lintel (the earliest in the town is 1610) and, on the left, note the attractive Horsemarket, with features like forestairs, harling, crowstep gables, pantiles and dated lintels. On a bit, facing the Stables Gallery is the youth hostel (SYHA) which may originally have housed the children brought in to slave in the weaving factory.

The street leads out to the central High Street – East Port, with Town Hall, Palace, Fountain, etc. Before exploring the palace look at the plaques on the houses opposite (Hunting Lodge Hotel) and it is probably worth buying one or two of the local guide books for the palace and town.

The Royal Palace (NTS) is the major attraction, the 16th century gatehouse and elegant Renaissance frontage dominating the town centre. It was the much-loved hunting seat of the Stewart monarchs, somewhere to escape the burden of state duty – a sort of early Balmoral. Its highlights are the Chapel Royal, the King's Room, the re-created gardens and the 1539 Royal Tennis Court, still in use.

Falkland Palace gateway.

The Bruce Fountain may be on the spot of the original town well. The heraldic lions hold shields with Bruce and town coats of arms, the latter a stag lying below an oak. The nearby statue is to Onesiphorus Tundall-Bruce, the local laird, who died in 1855. The shape of a cross on the road indicates the site of the town's mercat cross.

Walk up Cross Wynd and turn right along Brunton Street. The coat of arms over one door is of the hereditary royal falconer. At the end turn sharp right along the High Street back to the town centre. Falkland Weavers (good cafe) is on the right and, next to it, the improbably-sited local electricity sub-station. Cameron House was the likely birthplace of the famous covenanter whose name was taken by a famous regiment.

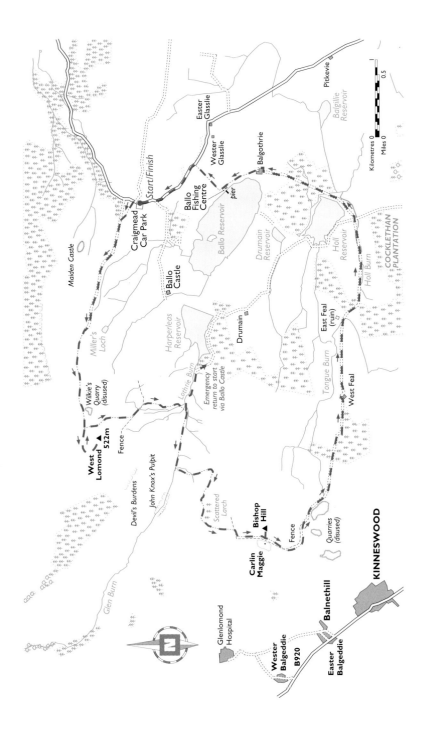

A LOMOND HILLS CIRCUIT

his is a major hill walk and, taking in the highest summit in Fife, West Lomond, 522 m, it is worth saving for a clear day. It should be treated as a proper mountain expedition, going well-equipped and supplied. (In summer carry water.)

The Lomonds bulk large in views of Fife (though, to be quite accurate, the Bishop is in Kinross-shire) and the high road from Leslie to Falkland which divides the range gives easy access and considerable height before starting: Craigmead car park is 285 m. From Falkland you simply drive along the High Street and the continuation is indicated. From the south (Glenrothes/Leslie) some care is needed as the turn-off in Leslie (a narrow road near the east end of the town) is difficult to see, though signposted for Lomond Hills/Falkland. Craigmead is obvious and has toilets, information/map board, and picnic tables.

The walk starts in the corner of the car park, signed for West Lomond, runs along a hollow and up steps. Turn left on the green paths which soon merge with a track coming up from the edge of a plantation, and follow the track over the moorland. (The Maiden Castle, a prehistoric fort, with very clear defensive ditches at its base, is well worth the detour for those interested in ancient sites.)

The big track gives easy walking. There is one stile and then, at the foot of the final cone, slope off round the hill to the right, a diversion which is slowly healing the scar of the path that once ran straight up to the summit. The path rises steadily and reels in an ever-changing view: Highlands, Ochils, far Ben Vorlich and Stuc a' Chroin, Loch Leven, the Cleish Hills . . . before finally reaching the white trig point and the black stones of a cairn. 'Brilliant!' was the response of one youngster.

Retrace the ascent route to the diversion notice at the foot of the final cone and continue round the foot of the hill, more or less south, then SSE, aiming

INFORMATION

Distance: 17½ km (11 miles).

Start and finish: Craigmead car park on the hill road over the Lomonds from Leslie to Falkland.

Terrain: Much on good paths and tracks but some rough hillsides and the walk should be treated as a mountain expedition, wearing boots and carrying appropriate gear. Take food and water.

Toilets: Craigmead car park.

Refreshments: None on route; Falkland afterwards (3 km).

eventually for some 'crumpled' ground, where several springs and streams start. There is a fence anyway and this is crossed at a stile, the Three Marches stile, just east of the broken ground. Descend the steep grass of the field, near the deep cut of the burn, and, when the level ground is reached, cross a stile and continue along the other side of the wall. There's a stile in the corner, which is on the ancient route from Glen Vale to Harperleas, and our path swings west to join this old path. (The infant Lothrie Burn could be the only water for drinking for the rest of the walk). If anything demands a shortening of the route, you can regain the start by going east via Harperleas Reservoir and the ruined Ballo Castle.

Glen Vale and John Knox's pulpit.

This wide col lies between West Lomond with its weathered exposures (the Devil's Burdens) and the wide flank of the Bishop. The track west is good and when it swings over to the wall (left) there is a gate, which you go through, but first continue for a few minutes to look down Glen Vale, a place worth exploring on another walk.

Once through the gate follow the track up the slope until a gate in the wall line is reached. Just before it, turn right, through the broken walling, and head west up into a landscape which looks like a carpet that has been rumpled into ridges and hollows, through which you pick your way (good sheep trods), and being surprised, too, by discovering a scattering of mature larch trees in this unlikely spot. Work through, up and leftish, till the path/overgrown wall along the scarp of the hill is reached. It is a dizzy drop on the other side and the prevailing west winds, hitting this barrier of hill, shoot up to give excellent lift to the gliders based at Portmoak (Loch Leven). You may well see and hear them overhead, and below!

Carlin Maggie – Bishop Hill.

After crossing a fence bear right to keep by the fence line as this gives a view of a pinnacle below the crags called Carlin Maggie (Carlin = witch). Tradition says Maggie had words with the Devil who was building West Lomond and he dropped his burden and rounded on Maggie, turning her to stone. The slim 'figure' was taller but a decade ago lost her 'head'. The pull beyond brings the summit bump (461 m) into view, a bit off-left, and marked by a cairn. Some would claim it as the best summit view in the Lomonds.

Return to the path along the fence on the scarp's edge. Pick out the path before leaving the cairn. It runs to a gate; you go through this and, beyond the path becomes a green track, joined by another which comes right up that steep west flank. It is worth peering over. The whole area hereabouts was once extensively quarried for limestone and the track wends along to pass several defunct quarry sites before reaching the larch plantation and the track down to West Feal. In 1852 there was a 'gold rush' with thousands camped here and the villages below were cleared out of all supplies.

To Arnot Reservoir from West Feal.

Timber extraction has rutted the old track badly but you can walk on the grass beside it. West Feal is no longer inhabited and East Feal, down the straight bit of track, is ruinous. A burn companions the road to some farm buildings and then you come to a crossroads and the tarred road that gives access to this hidden heart of the Lomonds.

Cross over to drop down below the dam of the Holl Reservoir and past the Italianesque-looking water treatment works. Beyond, after admiring West Lomond over the loch, go through a kissing gate and up the break in the plantation (also followed by a line of power poles). The crest above suddenly opens the view onto Ballo Reservoir, the largest of the Lomond six.

Descend to Balgothrie Farm, where diversionary arrows should be followed to avoid the farm buildings. Continue beyond and, when the Ballo Fishing Centre complex is reached, keep to the field edge behind it then, just beyond, turn up along the field edge and keep on climbing till a last, large stile deposits you on the Leslie-Falkland road. Turn left and a 15-minute tramp will lead back to the Craigmead car park.

INDEX

Other titles in this series

Long distance guides published by HMSO

Printed in Scotland for HMSO by CC. No. 70343 50C 4/95